Transactions
of the
American Philosophical Society
Held at Philadelphia
For Promoting Useful Knowledge
Volume 90, Pt. 3

Franklin's Father Josiah: Life of a Colonial Boston Tallow Chandler, 1657-1745

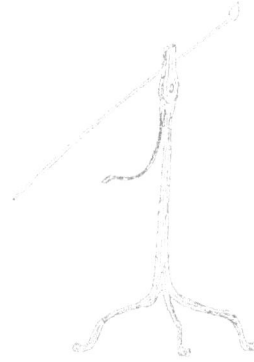

Nian-Sheng Huang

American Philosophical Society
Independence Square ⚖ Philadelphia
2000

ISBN:0-87169-903-6
US ISSN: 0065-9746

Library of Congress Cataloging-in-Publication Data

Huang, Nian-Sheng, *1951-*
    Franklin's father Josiah: life of a colonial Boston tallow chandler,
    1657-1745 / Nian-Sheng Huang.
       p.   cm. -(Transactions of the American Philosophical Society,
       ISSN 0065-9746 ; v. 90, pt. 3)
    Includes bibliographical references and index.
    ISBN 0-87169-903-6 (pbk.)
     1. Franklin, Josiah, 1657-1745. 2. Boston (Mass.)--History--Colonial
    period, ca.
    1600-1775.   3. Boston (Mass.)--Social life and customs--18th
          century. 4. Boston
    (Mass.)--Biography. 5. Candlemaking--Massachusetts--Boston--
    History-- 18th century. 6.
    Soap trade--Massachusetts--Boston--History-- 18th century. 7.
    Franklin,
    Benjamin,--1706-1790--Family. 8. Fathers--Massachusetts--Boston--
    Biography. I. Title.
    II.   Series.

F73.4.F73 H83 2000
974.4'6102--dc21
                                     00-030639

CONTENTS

List of Illustrations

Figure 26.   Tombstone of Josiah and Abiah Franklin and the
Franklin family grave in Granary Burial Ground behind
Boston Athenaeum.

(Except for Figure 6, all maps, drawings, and photographs
were made by the author.)

In Memory of My Parents

and

For Ching, Sha Sha, and Enid

Whose Warmth, Love, and Support Have Made This Work

Possible and Meaningful

❦ Introduction ❦

Josiah Franklin remains a marginal figure in most biographies of his well-known son, Benjamin Franklin. Because of a lack of written documentation, the essential information of his life changed little over the years beyond what the son recorded in his *Autobiography*. Biographers of Franklin included him mainly from a genealogical viewpoint, and few of them gave him further attention once the basic data of births, marriages, and deaths in the family were introduced. A tallow chandler and soapmaker for life, his unexciting career could not compare with his brother Benjamin Franklin the Elder, whose interest in poetry at least influenced Franklin's literary taste. The father's role in his son's life was also overshadowed by another son, James Franklin, printer and publisher, whose newspaper the *New-England Courant* provided rich materials of contemporary journalistic genres and public controversies for later generations to analyze.

Arthur Bernon Tourtellot's extensively researched *Benjamin Franklin*, published in 1977, greatly enhanced our understanding of the father. Compared with many biographers, he presented a much fuller picture of Josiah Franklin's personal experiences, especially his English background and lineage. Recently, I visited several archival sites in the Boston area and found some fragmented and yet valuable manuscripts concerning the father. Here I have reconstructed his life according to these findings, such as his bills, letters, subscriptions, participation in petitions, and court warrants for his legal disputes. I have also drawn information from newspapers, diaries, business accounts, inventories, deeds, and probate records which were useful to assess his trade and financial circumstances. What follows will not dramatically alter our previous perceptions of the father. But I do hope that this reconstructed life can help to appreciate an ordinary colonial, whose daily work, hardship, and frustration to maintain a large family should not be reduced to a nominal existence in the shadow of his son's fame.

The opportunity to search into his life satisfied my personal desire not only as a Franklinist but also as an Americanist in general. Since the time when I first read Carlo Ginzburg's classic *The Cheese and the Worms* (1980) in graduate school, I have been inspired to write a life of an ordinary person. I noticed, however, that although many historians were committed to a similar goal, their stories were often "problem" or "group" oriented. That is to say, a person's life could hardly catch the historian's attention unless first, he/she was recorded to have had some legal trouble or second, his/her story could serve as a mirror of the life and interest of a certain social group that the individual was believed to represent. These trends forced me to ask the question: if a person did not have a scandal or a legal trouble and if this individual's life had little to do with momentous events in the past or theoretical paradigms of the present, did that individual have a life worth recording? And if so, how can a historian approach it? Josiah Franklin, the case in point, missed many milestones cherished and magnified in conventional history. A Puritan himself, his parents nonetheless were not from the East Anglia Association, the heartland of the Puritan movement. Born after the Civil War, he had no chance to be associated with the first Great Migration beginning with John Winthrop and his followers. His experience of transplantation coincided with a period described in most history textbooks as the time of the Glorious Revolution, of the Salem witchcraft trials, and of Queen Anne's War, in none of which he was a player. He did not keep a diary, nor did public records reveal much trace about him. A neighbor to both Cotton Mather and Samuel Sewall, few historians would now give a second thought to these three as contemporaries in Boston. Other than the fact that he was Franklin's father, Josiah's life seemed to be too insignificant to tell us anything about our past. Yet was it not, I wondered, that his career was highly representative because his quiet and inconspicuous mode of life could have been shared by hundreds of whose early settlers who had no diary, no legal trouble, and no prominent public roles in their lifetime? I was therefore delighted when I found those fragmented pieces of documents suggesting how a colonial personality as marginal as a tallow chandler could have survived. I decided to write his life, believing that if historians do want to recapture the real lives of ordinary people, this individual, no matter how uneventful his life may appear to have been or how ambiguous his impact on the celebrated son still remains, deserves no less of our serious attention had he not been the father of Benjamin Franklin.

HOMELAND

🔖 "Josiah, my father, married young, and carried his Wife with
three Children unto New England"🔖

The ancestors of Josiah Franklin came from Ecton, a tiny village of
several hundred people four miles eastward of Northampton, England.[1]
The family used to have a small estate, aided by additional incomes
from a blacksmith shop. Born on December 23, 1657, Josiah was the
youngest son of Thomas Franklin 2nd (1598-1682) and his first wife
Jane White Franklin (1617-1662). "On the Wall of My Fathers par-
lour," one descendant recalled,

> Was Writen in Church-Text, Round about the Room, near
> the floor above it, the 16 and 17 verses of 3 John. God so
> Loved the World that He gave his only begotton son that
> Whosoever beleeveth in Him Should not persih but have
> Everlasting life. For God sent not his son into the world to
> condemn the world but that the world through him might be
> saved.[2]

The couple brought into this world eight sons and a daughter, and
Josiah was believed to resemble the father the most both to his person
and to his natural disposition. Dark skinned and corpulent, Thomas
was "of a chearful temper, pleasant conversation." Little hair on his

---

[1] I drew information for this segment primarily from Benjamin Franklin the Elder's *Short
Account of the Family of Thomas Franklin of Ecton in Northampton Shire, 21 June 1717*, see
Appendix, Document 13. Also useful are Arthur Bernon Tourtellot's *Benjamin Franklin:
The Shaping of Genius, The Boston Years* (Garden City, N.Y., 1977), chs. 1-5; Leonard
W. Labaree et al., eds., *Genealogy and Genealogical Charts, The Papers of Benjamin
Franklin* (New Haven, Conn., 1959), vol. 1, pp. xlix-lxxvii; Josiah Franklin to Benjamin
Franklin, May 26, 1739, *The Papers of Benjamin Franklin* (New Haven, Conn., 1960),
vol. 2, pp. 229-32; James Parton, *Life and Times of Benjamin Franklin* (1864; Boston,
1892), vol. 1, chs. 1-3; John W. Jordan, "Franklin as a Genealogist," *Pennsylvania
Magazine of History and Biography* 23 (1899): 1-22; John Cole, *The History and
Antiquities of Ecton* (1825; Philadelphia, 1865); and George Makepeace Towle,
"Franklin, The Boston Boy," in Justin Winsor, ed., *The Memorial History of Boston*
(Boston, 1882), vol. 2, pp. 269-96.

[2] Benjamin Franklin the Elder, *Notebooks*, vol. 1, p. 7, American Antiquarian Society.

head and always wearing a cap, he was an ingenious craftsman who not only was skillful in his own trade, but also tried his hand in the works of a clockmaker, a carpenter, a surgeon, and a gun-smith. Self-educated, he wrote handsomely and was a scrivener as well. His pious observances, good understanding, and diverse interests in history, astronomy, and chemistry made him a likable companion to the local gentry, including arch-deacon John Palmer.

Mother Jane, on the hand, was a tall and fair woman "Exact in her morals" and "Religiously Inclin'd." She kept a regular private meeting on Thursdays with her godly neighbors and would severely chide you if you were tardy in spiritual learnings. She suffered a long illness and died at age 47. Of her nine children, two infant twin-boys did not live long. Another son Samuel died at 23 (1641-1664), and a fourth son Joseph passed away at 37 (1646-1683). Twenty years older than Josiah, Thomas Franklin 3rd, the eldest son, learned the father's trade as a blacksmith and inherited the family farm. He also became a highly successful scrivener and would later accumulate an estate of £2000. Yet, his hot temper forced the father to leave the household about 1666 and to move to stay with his second son John, who had a dyeing business at Banbury in the neighboring Oxfordshire. Gentle and agreeable, John also took Josiah, then nine years old, as his apprentice.

A long established agricultural center, Banbury was situated at a busy crossroads on the west bank of Cherwell River in a deep valley of the northern part of the county, the famous rich red-land district. Less than thirty miles from Ecton along the Northampton Road to the northeast, Banbury (Figure 1) was also a town of growing populations and manufactures, recovering from some of the bloodiest battles of the region during the Civil War, such as the Battle at Edgehill and the Great Siege of the Castle.[3] The father-and-son relocation proved to be felicitous and brother John's kindness must have been paid off, because within ten years Josiah was able to finish his apprenticeship and marry Anne Child of Ecton in 1677, when he was not yet twenty years old.

The couple had their first daughter Elizabeth in March 1678,[4] and their

---

[3]A highly informative introduction to Banbury, its land, people, and history is in Alfred Beesley's, *The History of Banbury* (London [1841]).

[4]Josiah Franklin recalled that he lived with his brother John for eleven years. See his letter to Captain Benjamin Franklin, Blenheim, England, January 11, 1743/44, printed

Figure 1.   Banbury, c. 1730, based on a sketch in Alfred Beesley's *History of Banbury* (1841).

first son was born three years later in Banbury in 1681. Thomas Franklin 2nd died the next year in 1682, when John, at the age of forty-eight and after several failed courtships, finally married Ann Jeffs of Marton, Warwickshire. They were soon expecting a child in 1683 and so were the other couple Josiah and Anne. In the meantime, brother

in George Makepeace Towle, "Franklin, The Boston Boy," in Justin Winsor, ed., *The Memorial History of Boston* (Boston, 1882), vol. 2, pp. 270-71. He might have moved back for some time to Ecton where Elizabeth was born, *Genealogy* and *Genealogical Charts*, *The Papers of Benjamin Franklin* (New Haven, Conn., 1959), vol. 1, p. lvii.

Benjamin, also a dyer whose career was not going well in London, came to town and planned to marry Hannah Welles, the youngest daughter of the famous deceased dissenter the Reverend Samuel Welles, of Banbury in late 1683. Apparently, it was about this time that Josiah felt that three dyers from the Franklin family working in the same place were a little too crowded. The town had several other dyers and a much younger dyer Mr. Henry Clarson would later marry a daughter of the mayor and become a mayor himself in the future.[5] Tradition prevailed. After all, hadn't all his brothers yielded to primogeniture, learned other professions, and moved out of their hometown to London and elsewhere?[6] Rather than stay and compete with his two elder brothers, Josiah decided to leave.[7] He, Anne, and their three children, Elizabeth

[5]John Cross and Thomas Crossby were dyers at Banbury, see Banbury Historical Society, *Baptism and Burial Register of Banbury, Oxfordshire* (Oxford, England, 1966-69), Part One, p. 95; Part Two, p. 102. Henry Clarson married Abigail Harvey, daughter of Malachi Harvey (mayor 1709-10), in 1702; he held numerous public offices and became mayor for 1732-33, 1735-36, 1743-44, and 1747-48. Banbury Historical Society, *Banbury Corporation Records: Tudor and Stuart* (Oxford, England, 1977), pp. 303, 310.

[6]The second eldest son Samuel was apprenticed to Mr. Wilkinson, a silk weaver in London; John, the third son, to Mr. Glover, a cloth dyer in London and later settled at Banbury; the fourth son Joseph to Mr. Titcomb, a carpenter in London and later moved to Aldborough and Knatshah; the fifth son Benjamin to Mr. Pratt and Mr. Paine, skein silk dyers London and later, after his marriage, returned and settled in London. See Appendix, Document 13.

[7]Franklin wrote in his memoirs that the family left England for religious reasons. "The Conventicles having been forbidden by Law, and frequently disturbed, induced some considerable Men of his Acquaintance to remove to that Country, and he was prevail'd with to accompany them thither, where they expected to enjoy their Mode of Religion with Freedom," Leonard W. Labaree et al., eds., *The Autobiography of Benjamin Franklin* (New Haven, Conn., 1964), p. 51. Arthur Bernon Tourtellot has indicated that the main reason could be economic, *Benjamin Franklin* (Garden City, New York, 1977), ch. 4. His interpretation sounds plausible, even though that may not exclude the feelings of religious pressure as Benjamin Franklin the Elder wrote, "Things not succeeding there according to his mind, wth. the leave of his frinds and father he went to New England in the year 1683." See Appendix, Document 13. Differences in personal temperament of these brothers might have also played a role because another dissenter in the family Benjamin the Elder, who married a lead dissenting minister Samuel Welles's daughter, did not feel pressed to migrate. He moved back to London instead.

5, Samuel 2, and the infant Hannah born in May, left Banbury for New England in the summer of 1683. The couple never saw their homeland again.[8]

According to Alfred Beesley's *History of Banbury* (London [1841]), Samuel Welles, William Viscount Saye, Nathaniel Fiennes, John Fiennes, William Cobb, John Doyley, Joshua Srigge, Robert Wild, and Edward Gee were among the well-known dissenters of the region. If some people did leave for religious and political reasons, they probably would have some connection with this group. Unfortunately, I was unable to find out their exact names. Banbury also had an unusually high number of people named Vicars, Vickers, or Vivers. The few Vicars in Boston, such as Joseph and Roger Vicars, might have come from that region. But I have no proof.

[8]The Franklins' nostalgic feelings were expressed in Josiah Franklin's January 11, 1743/4 letter to Captain Benjamin Franklin, Blenheim, England, see George Makepeace Towle, "Franklin, The Boston Boy," in Justin Winsor, ed., *The Memorial History of Boston* (Boston, 1882), vol. 2, pp. 270-71n. "This is the Church Whose preacher I did fear/These are the Bells I did Delight to hear," Benjamin Franklin the Elder began with these lines in his poem "On Ecton" (1702), *Notebooks*, vol. 1, p. 102, American Antiquarian Society.

BOSTON

🎨 "This obscure Family of ours" 🎨

Half a century after settlement, Boston, named for old Boston in Lincolnshire, presented mixed blessings to the Franklins when they arrived in October.[9] Like many emigrants before them, it was a huge relief when they first sighted land after eight weeks on the high sea. As the vessel was piloted through a maze of islands in the Outer Harbor, slowly finding what bewildered early sailors called the "Lost Town,"[10] their anticipation heightened that the long and dangerous voyage would soon end. (Figure 2) Once the vessel passed the gaunt-looking Castle Island on the left-hand side and sailed into the Inner Harbor, Elizabeth and Samuel were amazed to discover so many boats, ships, and watercraft of all sorts floating in the bay that it was impossible to count them.[11] Looking at a strip of land straight ahead, both they and their

---

[9]Descriptions about landmarks, streets, and other topographical characteristics of Boston were based on information drawn from Caleb H. Snow, *History of Boston* (2nd ed.; Boston, 1828); Samuel G. Drake, *The History and Antiquities of Boston* (Boston, 1856); Samuel Adams Drake, *Old Landmarks and Historic Personages of Boston* (Boston, 1873); Justin Winsor, ed., *The Memorial History of Boston*, 4 vols., (Boston, 1881-82); Nathaniel B. Shurtleff, *A Topographical and Historical Descriptions of Boston* (3rd ed.; Boston, 1891); Samuel Barber, *Boston Common* (2nd ed.; Boston, 1916); Annie Haven Thwing, *The Crooked & Narrow Streets of the Town of Boston, 1630-1822* (Boston, 1920); Darrett B. Rutman, *Winthrop's Boston: A Portrait of a Puritan Town, 1630-1649* (1965; New York, 1972); Walter Muir Whitehill, *Boston: A Topographical History* (2nd ed.; Cambridge, Mass., 1968); and William F. Robinson, *Coastal New England: Its Life and Past* (Boston, 1983).

[10]Caleb H. Snow, *History of Boston* (2nd ed.; Boston, 1828), p. 82. Joshua Scottow used the same term in 1691 to deplore the moral condition of Boston, *Old Men's Tears for their Own Declension mixed with Fears of their and posterities falling off from New England's Primitive Constitution*, quoted in Bernard Bailyn, *The New England Merchants in the Seventeenth Century* (Cambridge, Mass., 1979), p. 123.

[11]When special envoy Edward Randolph, sent by the Lords of Trade, first arrived in Boston, he was also impressed by the magnitude of New England shipping and ship building industries, see *His Maᵗⁱ. Instructions to Mʳ. Randolph*, March 20, 1675/6 and Randolph's *Report to the Committee for Trade and Plantations*, October 12, 1676, in Robert Noxon Toppan, ed., *Edward Randolph*, 7 vol., (Boston, 1898-1909), vol. 2, pp. 199, 247-50.

8

parents noticed its distinct landmark of three hills: a huge pole for beacon light erected at the top of Beacon Hill (formerly Sentry or Centry Hill), to its right a windmill on Snow Hill (later Capp's Hill) and to its left a battery at the foot of Fort Hill. (People in Charlestown used to call Boston *Trimountain* because, from a north-northwesterly view, they saw three peaks of Mt. Vernon, Beacon Hill, and Cotton Hill.) Presently, they could see throngs of townsfolk gathering on the wharves, chatting, yelling, and anxiously waiting to barter with the arriving vessel for shipboard provisions. The unfamiliar noisy scene suddenly hit an unsettling thought that they now must make this place home.

Once landed, they began to have a better sense of the new country. (Figure 3) Situated on a narrow peninsula of under eight hundred acres of solid land (that is, the size of one and a half Wellesley College campus today),[12] the so-called largest urban center in English North America had a population of five to six thousand souls, which was but slightly bigger than that of Banbury. The fist-shaped isthmus, a mile or so between the widest points, stretched about two miles long, connecting with the mainland only through a narrow path called the Neck, where a Town Gate shut down every night for protection. A mile away, most of those eight to nine hundred cottages and dwelling houses were packed into two districts—the North End and the South End, linked by a government and business section where a Town House and a Town Dock lied at the center.

For better or worse, after finding a temporary shelter, the Franklins could finally sit down, count their finances, and discuss their future

---

[12]Estimation of the actual size of early Boston, varying from 600 to 1,000 acres, has been made difficult not only because of the passing of time, but also because of the numerous swamps, marshes, and quagmires it had in the past. I adopted here the number 783 acres used by Edwin M. Bacon, *Boston: A Guide Book to the City and Vicinity* (1903; Boston, 1922), p. 2. This figure was confirmed by my measurement based on the widely publicized and very accurate John Bonner's map of 1722, which turned out to be 800 acres. See also John Gorham Palfrey, *History of New England* (Boston, 1865), vol. 1, 360n; Samuel Adams Drake, *Old Land Marks and Historic Personages of Boston* (Boston, 1873), p. 8; and *Report of the Record Commissioners of the City of Boston* (Boston, 1886), vol. 10, p. 217.

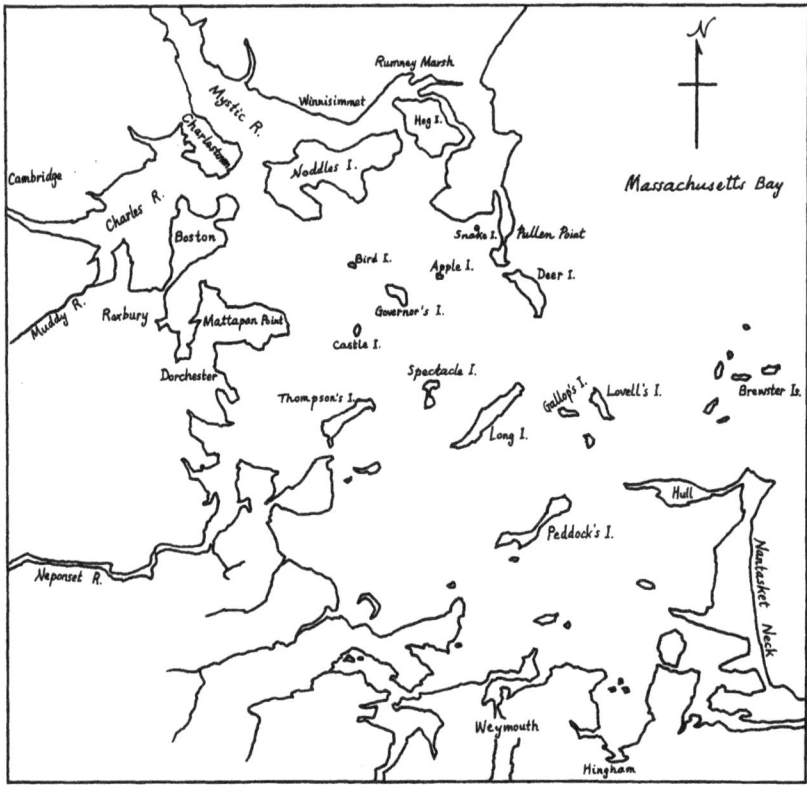

Figure 2.   Boston Harbor and vicinity.

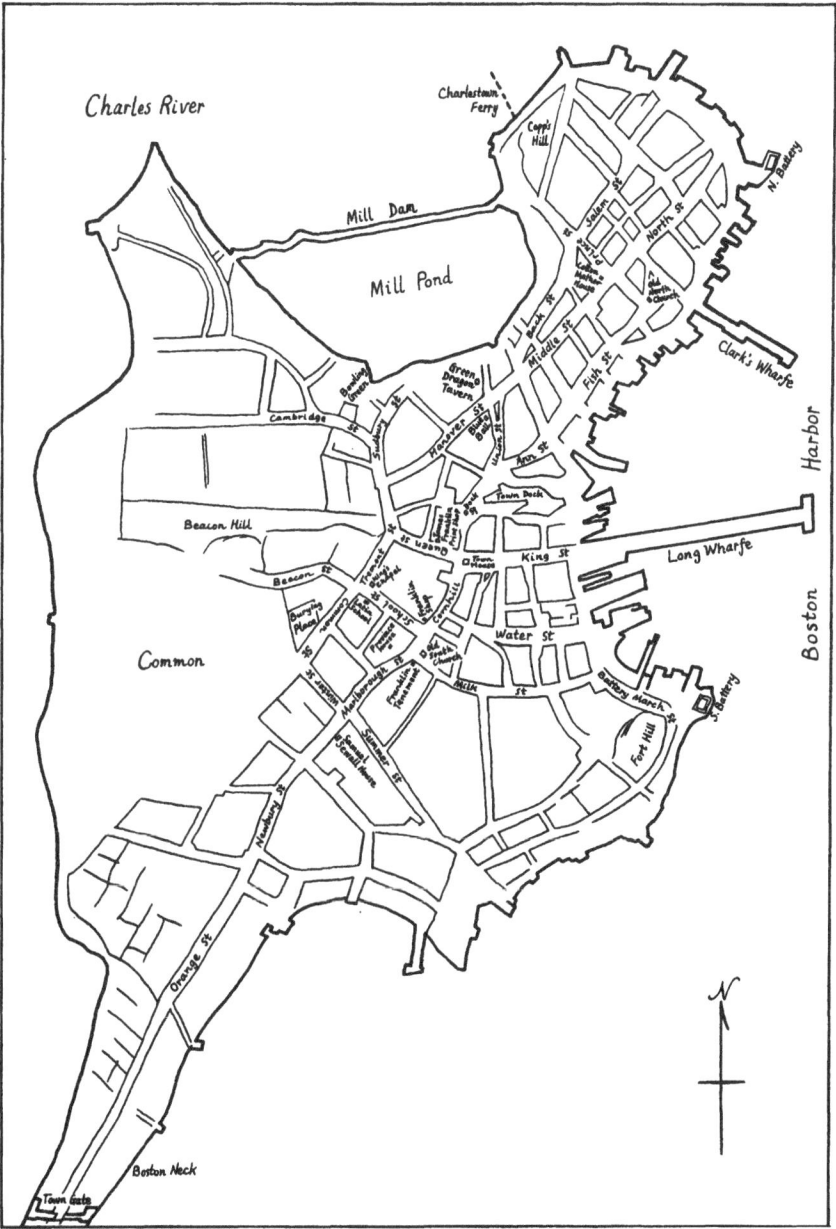

Figure 3. Town of Boston.

plans. The passage for two adults cost £10 and the three children £5.[13] A hefty bond of £100 must be prepared in order to pay the town after authorities examined their background and character and decided to admit them as residents.[14] An oath of fidelity was then taken, stating that "I will be true and faithful to this Government, & accordingly yield assistance thereunto with my person & Estate as in Equity I am bound." Curiously, the Franklins were directed to take the oath by holding up the hand, not by laying the hand on the Bible according to a standard practice in England. They were relieved, however, to be asked to pledge loyalty only to the provincial government, while the oaths of Allegiance and Supremacy were not required.[15]

Sometime during the first half of 1685 the Franklins rented a small house on Milk Street, right across from the Old South Meeting House.[16] The tenement, where the family would stay for twenty-seven

---

[13]Caleb H. Snow, *History of Boston* (2nd ed.; Boston, 1828), p. 10; William B. Weeden, *Economic and Social History of New England, 1620-1789* (Boston, 1890), vol. 2, p. 877; Arthur Bernon Tourtellot, *Benjamin Franklin* (Garden City, N.Y., 1977), pp. 50-51. A more recent assessment indicated that in the 1630s

> the cost of outfitting and moving a family of six across the ocean was reckoned at £50 for the poorest accommodation, or £60 to £80 for those who wished a few minimal comforts. A typical English yeoman had an annual income of perhaps £40 to £60. A husbandman counted himself lucky to earn a gross income of £20 a year, of which only about £3 or £4 cleared his expenses. Most ordinary families in England could not afford to come to Massachusetts.

David Hackett Fischer, *Albion's Seed: Four British Folkways in America* (New York, 1989), p. 28.

[14]They could put down the money themselves but the custom was to obtain two sureties to pay in their behalf. Years later, Josiah became one of the sureties for Brice Blare in 1722 and for Nathaniell Alcock in 1726. *Report of the Record Commissioners of the City of Boston* (Boston, 1885), vol. 13, pp. 108, 159-60.

[15]Edward Randolph, *Report to the Committee for Trade and Plantations*, October 12, 1676, quoted in Robert Noxon Toppan, ed., *Edward Randolph*, 7 vols., (Boston, 1898-1909), vol. 2, pp. 233-34. See also William Stevens Perry, ed., *Historical Collections Relating to the American Colonial Church* (1873; New York, 1969), vol. 3, pp. 6-7. Oath-taking formality was no small matter at the time; see Samuel Sewall's numerous references in M. Halsey Thomas, ed., *The Diary of Samuel Sewall* (New York, 1973), vol. 1, pp. 116-17, 133, 160, 163, 165-66, 543.

[16]The most useful and detailed description of the Milk Street dwelling house and property is Nathaniel B. Shurtleff, *A Topographical and Historical Description of Boston* (3rd ed.; Boston, 1891), pp. 615-25.

years, was built forty years earlier. Its original owner was a shoemaker named Robert Reinolds and his property, including a house and a garden, was valued at £110 in 1643. After Reinolds died in 1659 and his widow Mary Reinolds in 1669/70, their son Nathaniel Reinolds, also a shoemaker and a militia lieutenant, inherited the estate. A series of land deals in Boston and elsewhere threw Reinolds into financial strains and in 1691, he mortgaged the Milk Street house to Simeon Stoddard, Esq., for £50, but the Franklins stayed as tenants.

The size of a two-car garage in modern time, the tenement was a humble wooden structure measured at twenty feet square from the outside. (Figure 4) Clapboarded on the northerly side facing Milk Street, it had an attic with a pointed gable and two small windows on each side of the roof close to the back of the house. The door was on the westerly side with a narrow passageway in front of it. There were five small windows on the first floor: one on the westerly side toward

Figure 4. Franklin's tenement
in Milk Street, based on details
in Nathaniel B. Shurtleff's *Topographical
and Historical Description of Boston* (1891).
Previous drawings showed windows
too big to reflect the architectural
realities of the 1640s.

the end of the house, two on the northerly side facing Milk Street, and two on the easterly side. The second floor had the same number of windows at the same locations. Like most colonial houses of this type, the main floor had only one room no bigger than the standard size of sixteen to eighteen square feet. Here in a place smaller than a genteel parlor was the family's kitchen, dining room, and living room. All daily activities, cooking, eating, feeding, sewing, knitting, reading, praying, even Josiah's work, would take place here. In the back of the room on the south side was the fireplace, which connected to the chimney in the center of the house. Left to the fireplace there was a closet where the family kept all the kitchenware. To its right there was a small entry from the doorway and a stair leading to the second floor.

Most likely without any partition, the second floor had almost the same arrangement as the first. It was used as one big bedroom for the whole family, a practice quite common during this period when separate rooms were scarce and concerns for privacy were not an issue in average households. The size of the family, however, was growing fast. In the same year soon after moving into the tenement, the couple's first child in the New World, named Josiah, was born on August 23. Another daughter Anne came in 1687. The next year a third son, Joseph, was born and died within a week. After a second Joseph was born at the end of June 1689, mother Anne passed away on July 9 and so did the infant six days later. She had been married to Josiah for twelve years and given birth to seven children. She accompanied him from Banbury to Boston, and helped him and the whole family survive the most critical stage of the transatlantic migration and early settlement. Vigorous at the age of thirty-two but struggling alone to work and feed the large family day in and day out, the widower could not stay single for long. In less than five months later that year, he was lucky enough to find and marry Abiah Folger of Nantucket, then twenty-two,[17] who was sympathetic with him and felt sorry for the five motherless children. During the first ten years of their marriage, the new couple gave birth to five children in regular intervals: John in 1690, Peter 1692, Mary 1694, James 1697, and Sarah 1699.

[17]A year earlier, Abiah Folger received adult baptism at the South Church on August 19, 1688. The two therefore could have met at church activities. [Hamilton Andrews Hill and George Frederick Bigelow], comps., *An Historical Catalogue of the Old South Church, 1669-1882* (Boston, 1883), p. 303.

CRAFTSMAN

❧ "My Father . . . a Tallow Chandler and Soap Boiler. A Business he was not bred to, but had assumed on his Arrival" ❧

Keeping his fast growing family alive was all-important to Josiah but he soon learned that the Boston population had only a small appetite to patronize dyers of woolens, silk, and other delicate material.[18] An expensive luxury, silk in particular, was not viewed favorably by the Puritans. Massachusetts law specifically prohibited lower social ranks to wear it:

> that men or women of mean condition, should take upon them the garb of Gentlemen by wearing gold or silk lace, or buttons, or points at their knees, or to walke in great boots, or women of the same ranke, to wear silk or tyffany hoods, or scarfs, which though allowable to persons of greater estates, or more liberal education, yet wee cannot but judg it intollerable in persons of such like condition.

This law was first passed in 1651 and twice reaffirmed in 1662 and in 1672. Anyone convicted for the offense for the first time shall be admonished, for the second time fined for twenty-shillings, and the third time, forty shillings.[19]

It has been commonly assumed that because of such an unfavorable environment, Franklin gave up his profession as a dyer.[20] This assump-

---

[18]Brother John was a dyer of woolens and Benjamin the Elder of silk. Josiah, a journeyman working with John, was perhaps not yet specialized in the trade and therefore had to deal with all sorts of material from cloth, woolens to silk. But the general demand for the trade was low; see *The Autobiography of Benjamin Franklin* (New Haven, Conn., 1964), pp. 48, 53.

[19]*The Colonial Laws of Massachusetts* (1672; Boston, 1887), pp. 5-6. *The Laws and Liberties of Massachusetts, 1641-1691* (Facsimile ed.; Wilmington, Del., 1976), vol. 1, pp. 73, 168-69; vol. 2, pp. 231-32.

[20]Arthur Bernon Tourtellot, *Benjamin Franklin* (Garden City, N.Y., 1977), p. 52. Albert Henry Heusser indicated, in *The History of the Silk Dyeing Industry in the United States* (Paterson, N.J., 1927), that "so far as New England is concerned, early references to silk are few and far between," p. 93. For promoters of the silk industry see Jean

tion was only partially true. Massachusetts authorities frowned upon people of mean condition wearing silk, but never intended to bar the product from the colony across the board. Silk, as the law clearly stated, was "allowable to persons of greater estates, or more liberal education," who possessed silk of all sorts from early on. Silk stockings, lace, buttons, caps, barronets, undercoats, blankets, and even rugs were found in many inventories of substantial households.[21] Mrs. Ann Attwood of Plymouth Colony, for example, had as many as eighteen tablecloths and sixty-six napkins. Her "silk mohear petticoat" was worth nearly £2 at her death in 1654 when a man's fine suit or the best kind of a woman's petticoat would cost about 1s.10s.[22] Shopkeeper Bozone Allen of Boston had a full stock of silk merchandise as early as 1652, including "silk mohair at 3s.," "silk grogram at 7/6," "7 gr. Chain & other silk buttons at 34s.," "37 yds. Silk & silver lace at 5d." "9 doz. silk lace at 20d., "silk & gold fringe (per yd.) at 15s." and "Colored silk (per oz.) at 2s."[23]

It is also important to keep in mind that Massachusetts laws became silent about silk after the mid-1670s. The omission coincided with a growing presence after that decade of royal officials and their families, whose abundant silk-wear became an unmistaken symbol of wealth and social status that inspiring local families would have liked to imitate. Samuel Sewall took a special, if not entirely amicable, note of Sir Edmund Andros's laced scarlet coat when he first arrived as governor-general on December 20, 1686. Several others in his company were also in scarlet, he observed.[24] An inventory of King's Chapel, founded in 1689, revealed that its earliest possessions consisted of "six surplices of fine Bagg Holland," "a Communion Cloath of the finest Crimson Genoa Damask," "an old Communion Cloath of Silk

d'Homergue, with Peter Stephen DuPonceau, *Essays on American Silk* (Philadelphia, 1830).

[21]George Francis Dow, *Every Day Life in the Massachusetts Bay Colony* (1935; New York, 1967), pp. 242, 263, 266, 268, 271, 272, 273, 277, 279.

[22]John Demos, *A Little Commonwealth: Family Life in Plymouth Colony* (New York, 1970), pp. 37-38, 54-55.

[23]George Francis Dow, *Every Day Life in the Massachusetts Bay Colony* (1935; New York, 1967), p. 244.

[24]M. Halsey Thomas, ed., *The Diary of Samuel Sewall* (New York, 1973), vol. 1, p. 128.

Damask," and "two old Cushions of Silk Damask."[25] Years later, rich merchant Andrew Faneuil openly urged people to visit his warehouse in King Street, where he stocked "flowered Venetian Silks of the newest Fashion" imported "from London in the last ship."[26] The dyers' trade including silk dyeing, therefore, could be a useful profession, even though the demand was not widespread.

In fact, silk dyer Ambrose Vincent made a successful career in Boston about a decade later after Franklin's arrival. Not surprisingly, he was affiliated with Anglicans. Coming from England toward the end of the 1600s, he married Sarah Barber, sister of John Barber of Kipis, Yorkshire, in King's Chapel by assistant minister Christopher Bridge in 1703.[27] They had eight children, but only three daughters, Sarah, Hannah, and Jane, and a son Ambrose survived into maturity.[28] On the night of October 2 and 3, 1711, a fire broke out in Cornhill that forced numerous merchants and shopkeepers to evacuate.[29] Vincent was unfortunately one of the victims and gave the public this notice, "M[r]. *Ambrose Vincent* Silk Dyer and Scowrer that Liv'd in Corn-hill, *Boston*, before the late Fire, who dyes all manner of Silks, Cloth and Stuffs; and Scowres Mens Coats: Lives now on the South-side of Wings-lane, in the House Formerly Mr. Hills [Hulls]."[30] The hasty move also

---

[25]F. W. P. Greenwood, *History of King's Chapel* (Boston, 1833), p. 175. Many of those items were gifts of King William acting on his deceased queen's promise, and brought back from England by the Reverend Samuel Myles in 1696. André Mayer, *King's Chapel The First Century, 1686-1787* ([Boston] 1976), p. 11.

[26]*Boston Gazette*, January 25, 1719. Similarly, Mr. William Stoddard sold "New Fashion Silks, lately Imported" in Butler's Row, *Boston Gazette*, November 21, 1726.

[27]*Report of the Record Commissioners of the City of Boston* (Boston, 1898), vol. 28, p. 16.

[28]The Vincents had their first son Thomas on October 19, 1704, who died in less than a year. Two daughters, both named Mary (b. 1705/6 and 1708), died young. A third daughter Ann (1707) also died early. *Boston Transcript*, March 27, 1905, Note 697.

[29]One of the most devastating disasters in Boston up to that date, the fire lasted seven hours from about seven o'clock in the evening until two the next morning. Seven to eight people lost their lives and almost a hundred houses, along with the First Church and the Town House, burned. M. Halsey Thomas, ed., *The Diary of Samuel Sewall* (New York, 1973), vol. 2, p. 669, 669n. [Worthington Chauncey Ford, ed.], *Diary of Cotton Mather* (New York [1911]), vol. 2, p. 113.

[30]*Boston News-Letter*, November 5, December 10, 1711.

compelled him to dispose his over-stocked merchandise which revealed
an enormous inventory comprising "Broad Cloths, Irish Linnen,
Druggets, blew Holland and Linnens, Lawns, Indigo, Threads, Corks,
Small Arms, Hats, Fishing and Deep-Sea Lines, Marline, Sail, Twine,
Pictures and Prints."[31] Furthermore, a surviving receipt (Figure 5)
showed that his customers were not those who had small items like silk
buttons or lace, but those who could afford to order rich colors, guild,
lining, and various patterns of grounds and flowers dyed on their expensive coats, petticoats, and damasks.[32] Vincent was a clerk of the market
for 1716 and acquired the title of gentleman when he died in 1724.[33]

By then, the silk dyeing business seemed to have become a
booming enterprise with half a dozen migrant dyers rushing to Boston,
including Benjamin Franklin the Elder, silk dyer from London. (Figure
6) "This is to give notice, that there is lately arriv'd from England *George
Leason*, who with *Thomas Webber* of Boston, Cloathier, have set up a
Callendar-Mill and Dye-House in Cambridge-Street Boston near the
Bowling-Green . . . Dyes and scowers all sorts of silks."[34] Kindred
advertisements abounded and silk dyers such as James Vincent, Joseph
Herbert, Samuel Hall, Edward Carter, James Morris, John Quig, and
Rice Evens all came to town to promote their business.[35] In his

[31]Ibid., April 14, May 26, June 2, July 28, August 4 and 11, 1712.

[32]Ambrose Vincent's bill to Mr. David Stoddard, and to Madam Shrimpton, June 3,
1719, in *The Shrimpton Family Papers: 1630-1867, Reel 3: Miscellaneous Papers*,
Massachusetts Historical Society, see Appendix, Document, 14. Madam Elizabeth
Shrimpton was the wife of David Stoddard, son of Simeon Stoddard.

[33]Samuel G. Drake, *History and Antiquities of Boston* (Boston, 1856), p. 527. Drake
suggested that he died at 47. A contemporary Bostonian Jeremiah Bumstead called him
a "weighter" on March 30, 1724, see S. F. Haven, ed., *Diary of Jeremiah Bumstead of
Boston, 1722-1727, New England Historical and Genealogical Register*, vol. 15, p. 201.
Several Vincents later moved to a place in the South End, named Vincent Alley. It was
renamed Franklin Street when Charles Bulfinch built the first row houses in Boston in
1794.

[34]*Boston News-Letter*, April 28, 1712.

[35]For James Vincent, *Boston Gazette*, July 15, 1728, *Boston News-Letter*, July 18, 1728;
Joseph Herbert, *Boston News-Letter*, July 1, 1731; Samuel Hall, *Boston Gazette*, March
6 and 13, 1727; Edward Carter, *Boston Gazette*, April 22, 1728, June 30, July 7, 1735;
James Morris, *Boston Gazette*, October 31, 1737, January 23, February 6, 1738; John
Quig and Rice Evens, *Boston Gazette*, February 6 and 13, 1738.

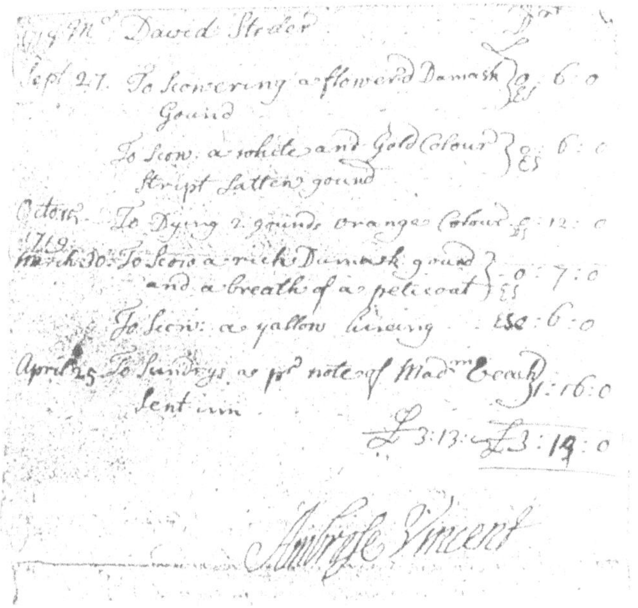

Figure 5. Ambrose Vincent's bill to David Stoddard, 1718-19.

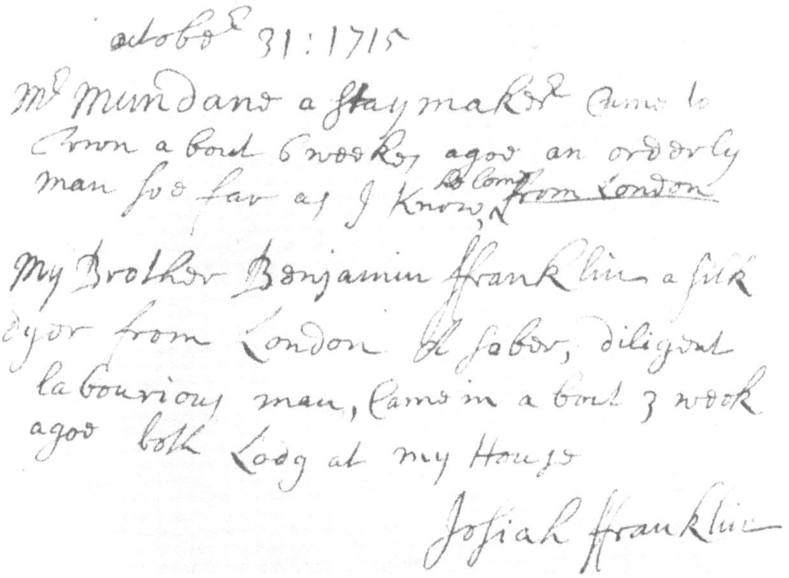

Figure 6. Josiah Franklin's deposition, October 31, 1715. Original size 9cm x 6cm. Photographed by the Boston Public Library.

announcement silk dyer Edward Carter was particularly mindful to point out his connections with the old Vincent,

> Edward Carter, Silk Dyer and Scowerer from London, who formerly kept the Rainbow & Blew Hand in Cambridge Street Boston; now keeps the Rainbow and Blew Hand in Wing's Lane, formerly Mr. Vincent's Dye-House, where is Dy'd and Scower'd all Sorts of Brocades, Velvets, Damasks, Sattins, Lystrings, Tabbies, Burdets, Mohairs, Poplins, Sasnets, Persions, Cloths, Camblets, Stuffs, Linnins, Needle-work and embroydery, Black Silks, White-Sasnet-Hoods, Fine Chince and Callacoes, Men's and Women's Silk & Worsted Hose, Beding, and all Sorts of furniture, New Clothes Scower'd Wet & Dry, all Sorts of Shop Goods Callendard, Prest and Pack'd for Sale; All for Ready Money.[36]

Such an influx of London dyers made the Boston market highly competitive, and few newspapers readers would fail to reckon their promotions as one of the most extensive and most costly to date. Adding to the fray were the occasional auctions of foreign spoils, which had the double-effect of arousing a popular curiosity in fanciful silk products and of dumping their prices at the same time, as this public pronouncement illustrated, "Sell captured ship *St. Francisco* & her Lading taken by Capt. Augustine Rouse of Her Majesty's ship *Saphire*: consisting all sorts of silks, gold & silver buttons & thread, silk stockings, Gold & Silver Lace, Ribbons, raw silk, linnen, hats," etc., etc.[37] Indeed, silk became such a favored commodity that it was frequently singled out as a key extravagance, which critics blamed to be the cause for a helpless outflow of silver money. One observer claimed,

> Many things have been *Imported*, which have *not been necessary*, yet very costly; such as *Silver and Gold Lace*, worn on *Cloaths* and *Shoes, Velvet, Rich Silk, Sattin, Silk Stockings, Fine Broad-Cloths, Camlets, Perriwiggs, Fine costly shoes and Pattoons, Ribbons, Rich Lace, Silk-Hankerchiefs, Fine Hatts, Gloves* of great price and little worth . . . that the Province

---

[36]*Boston Gazette*, April 22, April 29, and May 13, 1728; January 23, January 30, and February 6, 1738.

[37]*Boston News-Letter*, May 5, 1712.

Taking all these financial, material, and personal circumstances into account, Franklin made a prudent and necessary decision to change. The dyer's art did not entirely disappear from the family, however. Benjamin Franklin the Elder's notebooks contained detailed descriptions of the trade, including coloring, watering, dyeing, and scouring.[45] As dyer with more than fifty years of experience, he failed to become prosperous in Boston, perhaps mainly because of his age and poor health.[46] His namesake recalled many years later that mother Abiah was a very good dyer who had a "full & particular Receipt for Dying Worsted of that beautiful Red," which she passed onto her daughter Jane.[47] Even his brother James tried his hand in printing colored cloth

---

September 3 and wrote in his diary,

> Visited a Machine at Doctr. Franklin's (called a Mangle) for pressing, in place of Ironing, clothes from the wash. Which Machine from the facility with which it dispatches business is well calculated for Table cloths and such articles as have not pleats and irregular foldings and would be very useful in all large families.

John C. Fitzpatrick, ed., *The Diaries of George Washington, 1748-1799* (Boston, 1925), vol. 3, p. 235.

[45]Three of his notebooks or commonplace books survived. The first two volumes are in the American Antiquarian Society. The third, which has most of the information on silk dyeing, was published in *Publications of the Colonial Society of Massachusetts* (1906), vol. 10, pp. 190-225. For the technical aspect of silk dyeing see Thomas Cooper, *A Practical Treatise on Dyeing and Callicoe Printing* (Philadelphia, 1815), Charles O'Neill, *A Dictionary of Dyeing and Calico Printing* (Philadelphia, 1869), and Richard H. Gibson, *The American Dyer* (2nd ed.; Boston, 1878).

[46]According to Benjamin Franklin the Elder, there were four groups of skein silk dyers in London: Scarlet dyers who dyed only red in grain, light color dyer who dye all colors, heavy color dyers, and black dyers. "My practice Was upon Raw, that is, unwro't silk, in the skeyn, both black and Colours, for about Thirty years, and afterward I Dyed Garments of Wrought Silk and stuffs and cloth for about Nitneteen years More." *Commonplace Book, Publications of the Colonial Society of Massachusetts* (1906), vol. 10, pp. 222-23. He was sixty-five when he arrived in Boston in 1715 and subsequently suffered several illnesses. "Nov. 3, 1719. I came to dwel wth my son. On 18, had a shivering, on 19°, I fainted away. Aug. 22, 1725. Fainted at meeting. 1726 I have been much out of order this year with the Drosesie, & faintings." Benjamin Franklin the Eldler, *Notebooks*, vol. 2, n. p., American Antiquarian Society.

[47]Benjamin Franklin to Jane Mecom, July 17, 1771, in Carl Van Doren, ed., *The Letters of Benjamin Franklin and Jane Mecom* (Princeton, N.J., 1950), p. 128. Robert Campbell believed that the scarlet dyer was "the most ingenious and profitable Branch of the Dying Business." *The London Tradesman* (1747; Devon, England, 1969), p. 261.

in his shop, who claimed in the *Boston Gazette* that "the Printer hereby Prints Linens, Callicoes, Silks, &c. in good Figures, very lively and durable Colours, and without the offensive Smell which commonly attends the Linens Printed here." He did not mince words. When a year later someone attempted to forge himself, he immediately protested and threatened to bring such an intruder to justice,

> The Printer hereof having dispers'd Advertisements of his Printing Callicoes, & c. a certain Person in Charlstown, to rob him of the Benefit of said Advertisements and impose upon Strangers, calls himself by the Name of Franklin, having agreed with one in Queen Street Boston to take in his work. These are to desire him to be satisfyed with his proper Name, or he will be proceeded against according to Law.[48]

🦅 "He had a mechanical Genius" 🦅

After several failed attempts trying at different businesses,[49] Josiah began to reckon that candle and soap-making could be a practical and yet not saturated field, perhaps because of the tedious nature of the work and the small margin of profit. Undaunted, he made up his mind and set out to learn the new trade. A man of adaptation as well as great energy, action, and determination, he never turned back.

Dyers were very familiar with soap and most of them had a big copper, a principal investment for candle and soap-making.[50] But generally speaking, a good craftsman ought to have experience as much

---

[48] *Boston Gazette*, April 25, 1719; May 9, 1720.

[49] See Appendix, Document 13.

[50] The inventory and the executor's account of Josiah Franklin's estate, see Appendices Documents 23 and 24, revealed little information of what instrument he had as a candle and soap-maker. He did own a huge old copper valued at more than £16 and weighed 280 pounds, which suggested the possible scale of his production. Perhaps long before these accounts were taken, he had given some of his equipment to son John or Peter; both became candle-makers. As far as I know, no inventory or account of John's or Peter's estate survived. The average investment needed for a tallow chandler can be found in Thomas Clark's inventory, which shows that his instrument, worth about £22 total, included 120 candle rods, a candle mold and press, several metal kettles, and an old copper, see Appendix, Document 22.

as equipment.[51] Once essential materials such as potash, grease, and lime were obtained, a soap-maker could begin to make lye, a first step in the soap-making process.[52] He bored several holes on the bottom of a barrel, covered it with some straw, and put potash in it. The wood ash was then soaked wet but not dripping. After three to four days, he put the barrel atop several bricks laid around an empty pail, and began to pour a gallon of water every hour or two into the barrel and kept it dripping into the pail below. After some lye was drained from the ashes, he put lime (four quarts of unslacked or eight quarts of slacked stone lime for a barrel of soap) into two to three pails of boiling water, poured it into the ash barrel and let it run through. He should not pour the lime near the bottom of the barrel before any lye was drained,

---

[51]The following discussions on candle and soapmaking in this segment were based on Francis Bacon, *Sylva Sylvarum, or A Natural History in Ten Centuries* (9th ed.; London, 1670); Denis Diderot, ed. Charles Coulston Gillispie, *A Diderot Pictorial Encyclopedia of Trades and Industry* (New York, 1959); Mrs. [Lydia Maria] Child, *The American Frugal Housewife* (Boston, 1833); Gertrude Lincoln Stone and M. Grace Fickett, *Every Day Life in the Colonies* (Boston, 1905); Marion Nicholl Rawson, *Candle Days: The Story of Early American Arts and Implements* (New York, 1927); Ladislaus Edler Von Benesch, *Old Lamps of Central Europe and Other Lighting Devices*, trans. and ed. Leroy Thwing (1932; Rutland, Vt., 1963); L. M. A. Roy, *The Candle Book* (Brattleboro, Vt., 1938); Leroy Thwing, *Flickering Flames: A History of Domestic Lighting through the Ages* (Rutland, Vt., 1958); Ruth Monroe, *Kitchen Candlecrafting* (New York, 1970); Thelma R. Newman, *Creative Candlemaking* (New York, 1972); and Ray Shaw, *Candle Art: A Gallery of Candle Designs & How to Make Them* (New York, 1973).

[52]A common material, limestone was early found in Massachusetts, see Mr. Higgeson, "New-Englands Plantation" (London, 1630), in Massachusetts Historical Society, *The Founding of Massachusetts* (Boston, 1930), p. 84. There was a lime kiln near the Bowling Green (later Bowdoin Square) in Boston and John Blowers, mason, sold stone lime by the hogshead or bushel in School Street. See George Francis Dow, *Every Day Life in the Massachusetts Colony* (1935; New York, 1967), pp. 115, 134, 138.

An advertisement in the November 4, 1704 issue of the *Boston News-Letter* said that "there is lately set up at *Charlestown Ferry* in *Boston*, a Pottash-work, at the house of *John Russell Ferryman*, where all Persons that have any *Ashes* to spare, may receive *six pence* per. *Bushell* in *Money*, for any Quantity that they Shall Deliver at the Water-Side." Obtained from hardwood (such as fir, birch, and oak) ashes, potash was a major product in New England. European soap-makers valued this American export and graded it above those ashes from Russia, Hungary, and the Rhine. According to Philip Kurten, a soap and candlemaker at Cologne on the Rhine, to make 50 tons of soap he would need 1,840 lbs. of American potash, but 2,000 lbs. of Russian or Hungarian potash, or 2,300 lbs. of native Rhenish potash, *The Art of Manufacturing Soaps* (Philadelphia, 1854), p. 18.

because the lime would become like mortar and block the dripping of the lye. Finally, the color of ashes told him that enough lye was obtained and he could test the strength of the lye by dropping an egg into the pail. If it sunk, the lye was still weak and fresh ashes should be dripped. If the egg was buoyed up half way, the lye was too strong and water should be added. If the egg almost disappeared but a fingernail piece of its surface remained to be seen, the lye was perfect.

Two kinds of soap now could be made. To make cold soap, the soap-maker put two pounds of grease and fat into a pail of lye under the sun, stirred the pail daily, and the soap should come in less than a week. To make hot soap, he set a big kettle over a fire and poured three pounds of meat scraps, waste grease, and a pail of lye into it. While the kettle was bubbling briskly, he used a skimmer to take out bones, skin, and other impurities, and added a bit of quick-lime. When the liquid became thick as molasses, it meant the soap was coming. Soft as brown jelly, the hot soap was a thick ropy mixture and was often kept in a barrel at a cool place like the cellar.

The unassuming looks of household candles seemed simple enough but this could be deceiving. A good candle needed an appropriate type of wick, the heart of a candle. Bad wicks smoked or sometimes did not burn at all. If they did burn, they may have flickered and died in a short while. Skillful candle-makers knew the importance of the wick, its material, type, and texture as well as its right size proportioned to the thickness of a quality candle, which must have a smooth filtration and attraction through its wick to burn properly.[53]

Experience also taught that ropes, cords, flax, and linen were poor material for the wick, whereas cotton yarns and strings, especially four to six twisted threads could stand straight and make a good wick. A

---

[53]When a candle was lighted, the area close to the heat of the flame would begin to melt. The melted portion was absorbed by the wick and eventually consumed by the flame. If the wick was too small, it could not absorb the melted part quickly enough and the candle would drip. The flame could be drowned and extinguished by the excessive melting, or burn a well in the center of the candle, leaving a wasted outer shell. If the wick was too large, it would absorb the melted part faster than the flame could consume it and thus caused the candle to smoke. If fact, key to many ancient lighting devices was the wick, such as dipping a papyrus stalk in oil or soaking a rope in fat. Both were to prolong the burning of the wick. Candles emerged only after the purpose of the wick was changed into consuming the body of wax or fat, and really good ones evolved only when such consumption became perfected.

cotton wick did collect carbon soot at its top, which dimmed the light. So the soot must be snuffed from time to time. But before Cambacérès introduced the plaited or braided wicks in the 1820s, the twisted cotton wicks remained the dominant form. As a rule of thumb, a soft and fast-burning candle needed a thinner wick and a hard and slow-burning a thick one, such as a beeswax candle. Beeswax, however, was very expensive and hard to find in the colonies. Most colonials used animal fat to make candles. A slaughtered ox, for example, provided eighty pounds of suet, which could produce three hundred candles. Greasy and not entirely inexpensive, tallow candles were the best alternatives where beeswax was not easily available. Picky customers would further insist that only half bullock tallow and half sheep tallow could make a good bright candle. Sometimes, a bit of camphor was added if the candle-maker was really anxious to please a connoisseur.

In New England, wax myrtle, generally known as bayberry or candleberry, was a shrub with aromatic foliage and wax-coated berries which could be collected in the fall, boiled, skimmed, and used as a wax substance in making candles. Many liked them because they burnt "with a clear white flame, producing little smoke and emitting a most agreeable aromatic odour."[54] Paraffin, a derivative from petroleum and a popular material for modern candle-makers, did not come into usage until the 1850s. Similarly, the use of spermaceti, a crystalline substance from the head cavity of the sperm whale, in large-scale candle manufacturing was a new breakthrough starting during the second-half of the eighteenth-century.[55]

---

[54]Roy Genders, *Perfume Through the Ages* (New York, 1972), p. 187.

[55]Boasting a quality far superior than that of tallow candles, James Clemens advertised to sell spermaceti candles in 1748, see George Francis Dow, *Everyday Life in the Massachusetts Bay Colony* (1935; New York, 1967), pp. 127-28. In 1749, Benjamin Crabb (or Crab), petitioned the Massachusetts General Court for the sole privilege of making candles of coarse spermaceti oil. The request was granted but he moved to Rhode Island. Quaker merchant and entrepreneur Obadiah Brown hired him to conduct the business and was disappointed about the performance. He learned the trade secret by his own experiment. His candle factory at Tockwotton in Providence was completed in 1753 and the next year he was marketing the new candles in Boston. After his death in 1762, the Obadiah Brown and Company was reorganized as Nicholas Brown and Company, which became the leading manufacturer of spermaceti candles in the colonies. Wrapped in blue paper, the candles were three to the pound and weighted from 28 lbs. to 35 lbs. a box. Under increasing demand, the company shipped hundreds

Josiah Franklin always styled himself as tallow chandler and his principal way of making candles, like most people of his day, was dipping, carefully securing half a dozen equal-length cotton wicks onto a two-foot rod and placing weights at the end of each wick. Several rods may be prepared in the same fashion, depending on available material, space and anticipated time for work. Slowly lowering the first rod into a sizable kettle at least twelve to fourteen inches deep and full of hot tallow liquid, he would allow the rod to stay in the kettle for a while, but never for too long or the hot tallow would damage the wick. Next, he would withdraw the rod and allow the tallow to solidify around the wick by hanging the rod on a rack. Meanwhile, he would start to dip the second and then the third rod. The same process would repeat many times until the candles on all the rods reached the desired thickness. For a faster production, a candle-maker could dip two or three rods at a time either with twelve wicks if the candles were to be of six to the pound, or with sixteen wicks if the candles were to be of eight to the pound. No two dipped candles looked exactly the same even after they were treated on a heated glass or metal plate. But dipping remained the most common and inexpensive way to produce household candles in colonial days.

Special candles, however, required more skills and tools. Long and slender candles of eighteen inches or more, also called tapers, were made by pouring down wax from the top of lengthy wicks, which were fastened to a hoop that could adjust its height by a pulley. Thick cylinder candles or various shapes of block candles were usually cast in molds. Most colonial craftsmen used tin and pewter molds, but some had iron, brass, or wooden ones. Huge molds could have three sections of the neck, shaft, and foot, which may cast a single block candle of up to thirty and forty pounds. Mixed with at least fifty percent beeswax, a large candle for religious ceremonies must be made of the highest quality.[56]

---

of boxes of candles to Boston, Philadelphia, New York, and other British colonies in North America and West Indies. James B. Hedges, *The Browns of Providence Plantations: The Colonial Years* (1952; Providence, 1968), pp. 89-90. Carl Bridenbaugh, *The Colonial Craftsman* (1950; Chicago, 1964), pp. 112-13.

[56]The inside of the mold should be meticulously cleaned, dried, and thinly oiled. Only at the ideal melting temperature should the candle be cast. The mold was then carefully put into and secured in a pail of lukewarm water to let the candle to cool and solidify.

炁 "By constant labor and Industry" 炁

Recalling his youth, Benjamin Franklin wrote in his *Autobiography* that "I was employed in cutting Wick for the Candles, filling the Dipping Mold, and the molds for cast Candles, attending the Shop, going of Errands, &c. I dislike'd the Trade." Indeed, smelly, hot, and dirty, the endless toil of a tallow chandler and soap boiler would easily bore any ambitious youngster within days. Yet, compared with making soaps and candles in one's own shop, selling them involved a larger and more unpredictable external world. Of the many difficulties and uncertainties that the marketplace posed, the first was pricing. Traditionally, soap was sold either by the pound or by piece like a dozen or a hundred, and any piece of soap made by a reputable craftsman should weigh between 12 ounces and 1 pound. Prior to the Civil War, for example, a company of soap and candle-makers in Bristol regularly passed ordinances to dictate and adjust selling prices for its members. A base rate of soap at a penny for the pound was set so long as olive oil cost no more than £9 per ton. Thereafter, soap price would rise a penny whenever the unit price of a ton of olive oil increased for £3.[57] Thus regulated, the level of soap prices was maintained steady at 3d. a pound for twenty four years from 1604 to 1628, when a dozen soaps were sold at the consistent price of 2s. 6d. to 3s. Wholesale prices varied roughly within the same range because during the same period, the price of a hundred (also called the C) soaps fluctuated between 22s. and 26s., fifty soaps (a half C) between 11s. and 13s., and twenty-five (a quarter C) soaps between 5s. 6d. and 6s. 8d.[58]

Through these operations, any slight change of temperature or pressure would cause the candle to crack. After at least twenty-four hours, the candle could be taken out from the mold, which always was a very challenging and delicate task. Slowly but safely released from the mold, the candle must be treated spotless. A devout candle-maker would not even touch the surface and only handle the candle by its wick so that the dirt and oil on his finger may not mar his product. The finished ceremonial candle from a skilled craftsman's shop was no less than a work of art. The purity of its content, the richness of its color, the perfection of its design, the smoothness of its shape, and the sweetness of its natural fragrance, all made the bright, gentle, yet steady two-inch flame most pleasant and most holy to behold.

[57]H. E. Hott, ed., *The Deposition Book of Bristol, 1643-1647*, 2 vols., (Bristol, 1935-48), vol. 1, p. 229.

[58]Ibid., vol. 1, p. 234.

Candles, the company also decided, shall be sold at a penny per pound when tallow price was at twelve pence the stone (8 lbs.),[59] and the candle price would rise a farthing when the pound of tallow rose for every two pence. Therefore, when in 1607 the company set the price for buying ashes at 6d. per bushel and for tallow at 3s. 4d. per stone, it meant that a pound of candle would be sold at 3d. Reaffirming the importance of price regulation, each new price increase required that an oath be taken and signed by members of the guild. Participants also had to put up a pawn, ranging from a gold ring to silverware or a considerable amount of cash, to substantiate their pledge. Any violator would be fined and could forfeit their security valued between 40s. to £40, according to the seriousness of the offense. When reports indicated that violations persisted, the company selected four viewers or searchers who were authorized to check members' houses, shops, cellars, ware-houses, and workhouses to prevent abuses and corruptions.[60]

The company also ruled that the business of soapmaking could not mix with that of candle-making. Without authorization of the company, soap-makers could not make and sell candles, nor could candle-makers, soaps. All members must undergo appropriate apprenticeship, and their experience and workmanship must be authenticated by existing masters before they could enter the business. In order to ensure the quality of their products, members should take an oath stating that

> I do freely, & voluntarylie, without Constrainte of my owne accorde depose & sweare, that nether I my selfe, nor any Servante or servants of myne, Directly or indyrectly, or any other personne or personnes by or with my privitie or Consent, Knowledge, or approbation, in the makeinge & boylinge of blacke sope, commonly called Bristoll Sope, for me, or my use, shall or will, untill the first day or march next in suinge the date hereof, use any oyle whatsoever but only oyle olyve, unless the greater part of our Companie do att oure Common hall, resoulve & agree to the Contrary. So god me help.[61]

---

[59]Ibid., vol. 1, p. 232.

[60]Ibid., vol. 1, pp. 125-27, 219.

[61]Ibid., vol. 1, pp. 161-62.

In 1682, Boston town authorities also attempted to regulate candle-makers by announcing that "Candles made up for sale shall be under the cognisance of the Clarkes the market & be liable to be weighed & forfeited for want of being full weight as butter & bread is."[62] Yet little evidence suggests that candle and soap-makers were continuously subject to regulations, even though many other producers were, such as bakers, shoemakers, and coopers.[63] The obstacle that the new chandler Franklin faced came from a more deeply rooted tradition: namely, that many colonials made their own candles and soaps at home and therefore did not need to buy them from others. Who would be his customers if he was to survive?

Other than the Bible and some religious tracts, average families in Boston had little printed material to read after dark. Like sunset, candlelight signaled the time to finish business and not a few preferred to rise early in the morning to read in order to save candles.[64] Even if they did stay long into the evening, many would be satisfied with candlewood light or inexpensive home-made candles.[65] "Artificial light

---

[62]Quoted in M. Halsey Thomas, ed., *The Diary of Samuel Sewall* (New York, 1973), vol. 1, p. 55.

[63]See *The Laws and Liberties of Massachusetts, 1641-1691* (Facsimile ed.; Wilmington, Del., 1976), vol. 1, pp. 39-40, 85, 118, 250; vol. 2, p. 314.

[64]M. Halsey Thomas ed., *The Diary of Samuel Sewall* (New York, 1973), vol. 1, p. 360; vol. 2, pp. 733, 900. Another diarist William Byrd II recorded that he often rose at 5 or 6 o'clock to read, Louis B. Wright and Marion Tinling, eds., *The Great American Gentleman, William Byrd of Westover in Virgia: His Secret Diary for the Years, 1709-1712* (New York, 1963), passim.

[65]John Wintrop, Jr., in "Of the Manner of Making Tar and Pitch in New England" (1662), described that

> those same [pine] knots the planters split out into small shivers, about the thicknese of a figure, or thinner; and those they burn instead of candles, giving a very good light, and they call it candle-wood, and it is much used in New England and Virginia, and among the Dutch planters in the villages. But because it is something offensive, by the much sullying smoke that comes from it, they usually burn it in the chimney- corner upon a flat stone or iron, except occasionally a single stick in their hand, as there is need of light to go about the house.

Thomas Birch, *The History of the Royal Society of London* (1756-57; New York, 1968), vol. 1, p. 101. Another early observer William Wood insisted, however, that

> out of these Pines is gotton the candlewood that is so much spoken of, which may serve for a shift amongst poore folks; but I cannot commend it for singular good, because it is something sluttish, dropping a pitchie kinde

was dear, and of poor quality,"
Oxford economist James E. Thorold
Rogers wrote, "This fact goes far to
explain the very early habits of our
forefathers, and in college life the use
of the Common Hall."[66] Another
substitute for regular candles was
rushlight, (Figure 7)

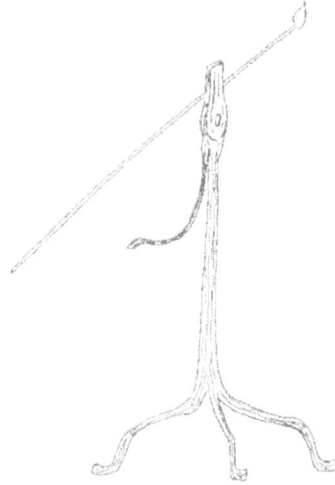

Figure 7. Rushlight.

sometimes also called rush candles,
which could be found in many
ordinary households.[67] "My grandmother," English essayist William
Cobbett recalled as late as the early-nineteenth century, "who lived to
be pretty nearly ninety, never, I believe, burnt a candle in her house in
her life." She cut the meadow-rushes when they grew substantial, that
is when each had a body of pith with a green skin on it. Then both ends
of the rush were cut off, leaving the prime part, which, on average, may
have been about a foot and a half long. The green skin was taken off ,
except for about a fifth part around the pith. This little strip of skin
would hold the pith together all the way. After soaking the prepared
rush in melted grease, and it was then put on an iron stand with a pair
of pliers to hold it. This rushlight cost almost nothing to produce and
was believed to give a better light than some poorly dipped candles.[68]

-----

of substance where it stands,
see his *New Englands Prospect* (1634; Boston, 1967), p. 19. See also Arthur H.
Hayward, *Colonial Lighting* (Boston, 1923) and Helen Brigham Hebard, *Early Lighting
in New England, 1620-1861* (Rutland, Vt., 1965).

[66]James E. Thorold Rogers, comp., *A History of Agriculture and Prices in England* (1887;
Vaduz, Japan, 1963), vol. 5, p. 382. See Appendix, Document 28.

[67]James E. Thorold Rogers wrote, "Rushlights were the commonest form of candle, and
of these rushlights, those which had but a slight coating of tallow were ordinarily used."
A History of Agriculture and Prices in England (1887; Vaduz, Japan, 1963), vol. 5, p.
381.

[68]William Cobbett, *Cottage Economy* (1822; New York, 1979), pp. 144-45. According

It seems that the first group of Franklin's customers must have been those who needed good illumination at night and clean linen and clothing on a regular basis. Persons frequently read and worked in the evening would likely buy candles if they could afford it. Taking both affordability and necessity into consideration, they tended to be those whose business activities would not end at sunset, such as merchants, shopowners, clerks, tavern-keepers, and shoemakers. In fact, because many shoemakers had to work through long winter nights, a special type of double candle, called shoemakers' candle, was invented for them. Composed of two molded candles joined together, it gave a strong and long-lasting light.[69]

Diligent ministers, scholars, and magistrates who often worked late into the evening needed a good supply of candles. A minister estimated that he needed at least three small candles (nine to the pound) every night, which cost 6 pence. If the yearly expenditure for candles of this frugal household could be at least £9 2s.,[70] those several dozen ministers and their families in Boston may constitute a significant group of clients. The work ethic of the Puritan divine Cotton Mather (1663-1728) was legendary. In addition to preaching four hours every Sabbath, he published 388 works in his life and 80 posthumously as a result of his constant labor in his study, where a warning sign to visitors simply stated: BE BRIEF.[71] When the ten-year old Benjamin Franklin was running errands for his father, he was likely to have delivered candles to the Mather's house, which was only four blocks away from home.[72] John Leverett, president of Harvard College, once bought close

---

to James E. Thorold Rogers, rushlights were sold at 5 ½ d. a pound in 1746 and a high priced "double rush candle" was in use until the late nineteenth century, *A History of Agriculture and Prices in England* (1887; Vaduz, Japan, 1963), vol. 7, p. 315, vol. 5, p. 382. The word rushlight was used by authors from Shakespeare and Milton to Dickens, see the *Oxford English Dictionary*.

[69]See *Candle, A Supplement to Mr. [Ephraim] Chambers's Cyclopaedia; or Universal Dictionary of Arts and Sciences* (London, 1753), vol. I, n. p.

[70]*Boston Gazette*, May 16, 1737.

[71]Kenneth Silverman, *The Life and Times of Cotton Mather* (New York, 1985), pp. 422-23; M. Halsey Thomas, ed., *The Diary of Samuel Sewall, 1647-1729*, vol. 1, p. vii.

[72]Benjamin Franklin's two letters to Samuel Mather on July 7, 1773 and on May 12, 1784 clearly said that he had been to the houses of both Increase and Cotton Mather.

to a hundred candles from Josiah Franklin, who was a regular supplier to him and his students. Candle supplies became such an integral part of the scholar's life at the college that President Benjamin Wadsworth charged a regular fee of candles for those who took room and board with his family.[73]

Related but not exclusive to those hardworking tradesmen and learned scholars, a second group of customers may have been those substantial families and sizable business establishments whose routine household keeping and business expenses included candles and soaps. These households usually had too many rooms to illuminate. Home-made candles would hardly be enough, as indicated by the large number of candlesticks often found on their premises. For example, when merchant William Paine died, the inventory of his estate showed that he had at least six brass and one silver candlesticks in the dwelling house, and as much as £1 10s. worth of wax candles in the warehouse. An inventory of the rich George Corwin, of neighboring Salem, also revealed that whereas the maid's chamber had no candlestick, there were a great candlestick in the hall, two iron candlesticks in the kitchen closet, one candle-box in the garret,[74] five candlesticks in the shop, and forty-four pounds of soap in the cellar.[75]

In most cases, these well-to-do families also possessed dresses, linens, cloth, curtains, beddings, napkins, towels of all sorts, which had to be washed from time to time. Some businesses had a similar

---

His familiarity with the internal structure of the latter's house gave the impression that he had been there several times. J. A. Leo Lemay, ed., *Benjamin Franklin: Writings* (New York, 1987), pp. 880-83, 1092-93.

[73]"George Eveleth from South Carolina aged abt 14. Years, was put here to board & goe to School, by Mr Jacob Wendal of Boston, merchant, who undertook to pay for his Board 14^s· pr week Board, 20^s· Washing & mending pr Quarter, & 5^s· Qrter for candles." Benjamin Wadsworth, *Diary* (1693-1737), May 15, 1733, Massachusetts Historical Society. See also his entries on June 13, 1715, January 28, 1733/4, November 23, 1734, April 11, 1735, and June 14, 1736.

[74]Francis Bacon recorded that "good housewives bury their candles in flour, or bran, which 'tis said increase their lasting, almost half." *Sylva Sylvarum, or A Natural History in Ten Centuries* (9th ed.; London, 1670), Century IV, Section 369.

[75]*Copy of Inventory of the Estate of Wm. Paine of Boston*, *Inventory of the Estate of Capt. George Corwin of Salem*, in George Francis Dow, *Every Day Life in the Massachusetts Bay Colony* (1935; New York, 1967), pp. 258-61, 270-83.

situation. Tavern-owners and innkeepers in particular would buy candles and soaps wholesale if they intended to have their establishments well illuminated and cleanly furnished. Thomas Selby, proprietor of the Crown Coffee House on Long Wharf from 1714 to 1725, fancied his premises with the latest English fashion and owned as many as thirty-eight table cloths, which must be washed clean to please his genteel patrons. At a time when most families in rural areas had only one or two candlesticks,[76] tavern-keeper John Turner of Boston had 15, Simon Rogers 19, and an owner of the Sun Tavern 17. Historian David W. Conroy described, therefore, that "evening was a period when labor could not be performed well in ordinary households because of their limits and cost of candles. In contrast, taverns were inviting beacons of light."[77]

A third group, shipmasters and traveling merchants, was frequently away from land for weeks and months when a good and reliable artificial light on board became necessary. Once a ship landed, sailors would seize the opportunity to enjoy themselves at local restaurants and taverns well into the night. Barbers and maid-washers welcome these transients who brought them business. Cleaned and refreshed, ship captains were often invited to gentlemen's residence, where dozens of men and women would drink, dine, sing, and dance until 2, 3, or 5 o'clock in the morning.[78] An increasing consumption of candles in this sort of entertainment pointed to the emergence of nightlife in Boston.

More important than personal and domestic indulgence, outgoing ships often ordered many boxes of candles and soaps for other colonies and overseas markets.[79] Candles, soaps, tallow, wax, as well as candle

---

[76]A good starting point to learn about the lighting condition in rural areas is Abbott Lowell Cummings, *Rural Household Inventories: Establishing the Names, Uses and Furnishings of Rooms in the Colonial New England Home, 1675-1725* (Boston, 1964).

[77]David W. Conroy, *In Public Houses: Drink & The Revolution of Authority in Colonial Massachusetts* (Chapel Hill, N.C., 1995), pp. 88, 89, 92.

[78]See "Extracts from Capt. Francis Goelet's Journal, Relative to Boston, Salem and Marblehead, &c., 1746-1750," *New England Historical Genealogical Register*, vol. 24, pp. 50-63.

[79]The export of candles became such a trend that Massachusetts lawmakers tried several times, in 1720, 1722, and 1724, to regulate it but to no avail. *Journals of the House of*

plates and sticks were import and export commodities subject to Parliament regulations.[80] Under the protection of navigation laws, however, they were not enumerated articles and therefore New England ships could carry them to any colony within the empire without paying an export duty.[81] South colonies such as Virginia and the Carolinas and the imperial outposts on Bermuda, Barbados, Nevis, St. Christopher, Bahamas, and Antigue, all imported candles, while ships leaving Boston frequently visited these places. In 1714, a small ship *Brunswick* of 65 tons left Boston for Barbados. It had a crew of ten men but carried 37 hogsheads of fish, 50 boxes of candles, and 15 boxes of soap.[82] Captain Charmian Betty instructed his agent to get beeswax and "40 to 50 boxes of tallow Candle of 8 to the pound," if coming from Boston or Rhode Island to Barbados, or to get "as much beeswax as possible, which was the best article," if coming from Virginia.[83] Because Boston was such a busy seaport,[84] selling wholesale to those incoming or

Representatives of Massachusetts ([Boston] 1923), vol. 2, p. 187; vol. 4, pp. 127, 128; and vol. 5, pp. 121, 126, 135.

[80]*Manufacturers and Other Products Listed in the Rates on Imports and Exports Established By the House of Parliament*, June 24, 1660, in George Francis Dow, *Every Day Life in Massachusetts Bay Colony* (1935; New York, 1967), pp. 246-57.

[81]Lawrence A. Harper, *The English Navigation Laws* (New York, 1939), pp. 396-97. This did not preclude local levies. By an act of 1722/3 (renewed in 1731), South Carolina, for example, charged these import duties: candles, 10s. @ cwt.; tallow, 7s. 6d. @ cwt.; beeswax, 7s. 6d. @cwt.; spermaceti, 2s. 6d. @ lb.; and spermaceti oil, 2s. 6d. @ gal., *Boston Gazette*, March 14, 1737.

[82]George Francis Dow, *Every Day Life in Massachusetts Bay Colony* (1935; New York, 1967), p. 155.

[83]William B. Weeden, *Economic and Social History of New England*, vol. 2, pp. 904, 905, 906.

[84]Bernard Bailyn and Lotte Bailyn, *Massachusetts Shipping, 1697-1714: A Statistical Study* (Cambridge, Mass., 1959); Robert G. Albion et al., *New England and the Sea* (Middletown, Conn., 1972); Bernard Bailyn, *The New England Merchants in the Seventeenth Century*. According to the reported figures in the *Boston News-Letters*, during the four weeks between October 22 and November 19, 1705, a total of 87 incoming and outgoing ships cleared from Boston, and during the five weeks from October 7 to November 11, 1706, 129 ships. In the last six months of 1714, as many as 236 ships had their clearances from Boston, see George Francis Dow, *Every Day Life in Massachusetts Bay Colony* (1935; New York, 1967), pp. 154-57. In 1748, 430 vessels came to and 500 vessels left Boston, see *Reports of the Record Commissioners of the City*

outgoing ships would have been crucial to the existence of Franklin's business. He did not need to look far for buyers because Samuel Sewall, his fellow parishioner, would load candles on a merchant ship for sale at Port Royal of Jamaica as early as 1689. (Figure 8)

Finally, institutional customers, such as schools, churches, and public facilities like meeting halls, court rooms, and office chambers were equally important. To celebrate Queen Anne's birthday on February 6, 1704/5, the *Boston News-Letter* reported that official parties lasted way into the evening, "the *Town-House*, and several Gentlemens

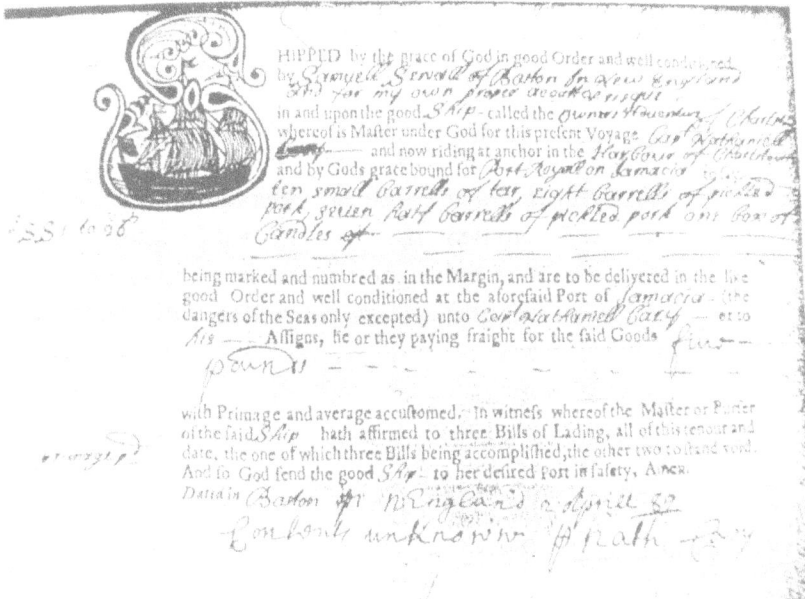

Figure 8.   Samuel Sewall's bill of lading, April 2, 1689.

---

of Boston, vol. 10, p. 219.

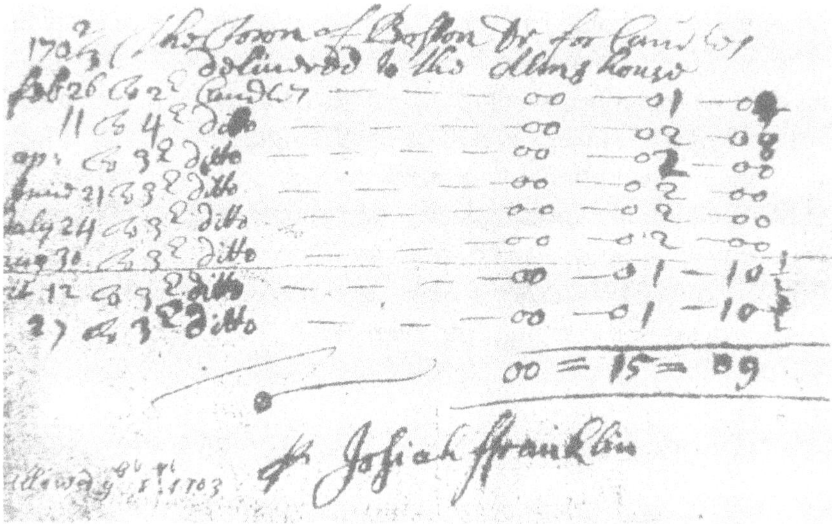

Figure 9.   Josiah Franklin's bill to Boston for candles to the Alms House, 1702/3.

Houses at night being full of illuminations."[85] Several years later, the same newspaper described that after officials celebrated King George's birthday at a noble enter-tainment inside the Council Chamber on May 28, 1716, "the Gentlemen met again in the Evening, where they repeated all the former Healths, which concluded with such extra-ordinary illuminations as were never seen in these Parts." In fact, of the very few surviving docu-ments concerning Franklin, several involved either with the magistrate's payment to him or with his billings to town authorities. (Figure 9) On March 4, 1703, Boston Committee for the Fortifying of Castle Island paid him 15s. for his tallow. On October 23 the same year, the same committee paid 3s. 9d. for his candles.[86] A few

---

[85]Such activity did not necessarily please every participant. Samuel Sewall complained that another celebration of the Queen's birthday in 1710/11 cost 5s. apiece. See M. Halsey Thomas, ed., *The Diary of Samuel Sewall* (New York, 1973), vol. 2, p. 632.

[86]Massachusetts Archives, 244: 31, 57.

years later in 1709, he charged the town for a total of £1 candles that he provided to the town watch between February 23, 1708 and April 23, 1709;[87] separately for 1710 and 1711, he sent a bill of more than £3 10s. for providing a full year of candles to the Batteries. (Figure 10)[88] Clearly, he had been a consistent provider to any local institution needing his products.

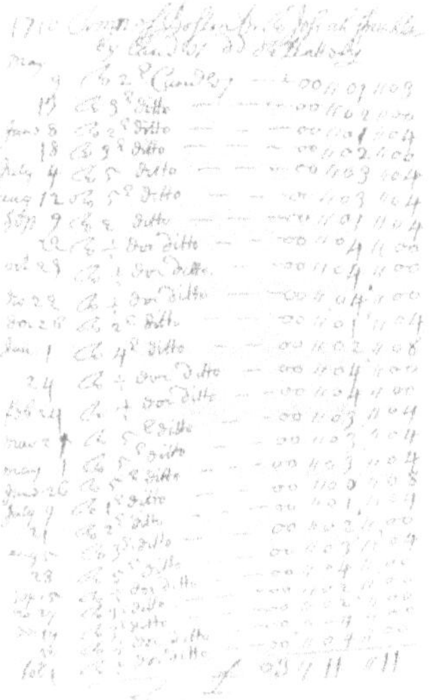

Figure 10.   Josiah Franklin's bill to Boston for candles to the Battery, 1710.

Franklin's business quickly passed the initial stage of survival. In 1691, he applied for permission to build an eight by eight foot "lean-to" at the back of his tenement house, which was approved at a town meeting on April 27.[89] A small but much needed storage place, the shed certainly would help the ever growing family, now of six children

---

[87]See Josiah Franklin's bill in George Makepeace Towle, "Franklin, The Boston Boy," in Justin Winsor, ed., *The Memorial History of Boston* (Boston, 1882), vol. 2, p. 271n.

[88]See Appendices, Documents 8 and 9.

[89]"Granted Libertie to Josiah Frankline to erect a buildinge of 8 Foote square vpon the Land belonginge to L' Nath' Reynolds neere the South Meeting house." *Report of the Record Commissioners of the City of Boston* (Boston, 1881), vol. 7, p. 207.

and one on the way. It could also be an indication that the demand for his candles and soaps was growing, and that he needed more work space to keep up with the business. Gradual improvement of the outlook of his trade gave him further confidence when a few years later in 1698, he ordered an iron sign twelve inches in diameter. He hung the sign by an iron bar outside of his residence, and townsfolk called it "The Blue Ball."(Figure 11)[90]

Figure 11.   The Blue Ball.

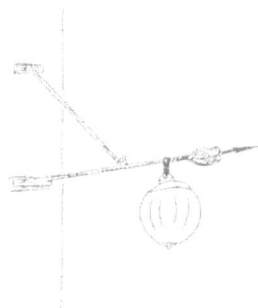

The public display of his business symbol showed his commitment to the trade and his hope for expansion. Most scholars believe that the Blue Ball was first hung on the Milk Street tenement and later at the corner of Hanover and Union Streets in 1712 when the family moved to a new house. In fact, Franklin managed to open a separate shop where he could concentrate on his work. It was a frontal portion of a house located close to the corner of School Street and Cornhill,[91] which

---

[90]Nathaniel B. Shurtleff, *A Topographical and Historical Description of Boston* (3rd ed.; Boston, 1891), pp. 624, 626-38; Samuel Adams Drake, *Old Landmarks and Historic Personages of Boston* (Boston 1873), p. 146; George Makepeace Towle, "Franklin, The Boston Boy," in Justin Winsor, ed., *The Memorial History of Boston* (Boston, 1882), vol. 2, p. 273.

[91]*Suffolk Deeds*, 23: 147. See also extracts in Thomas Minns, *Publications of the Colonial Society of Massachusetts* (1906), vol. 10, pp. 243-46. From the early 1640s and for much of the seventeenth century, this section of the main street was named High Street, as stated in the deed. It was changed into Cornhill around the turn of the century and remained so long after the Revolution until the name of Washington Street was extended to cover the entire main street. Nathaniel B. Shurtleff, *A Topographical and Historical Description of Boston* (3rd ed.; Boston, 1891), p. 616. Annie Haven Thwing, *The Crooked & Narrow Streets of the Town of Boston, 1630-1822* (Boston, 1920), pp. 117-18. This should not be confused with another High Street, named in 1803 to substitute for Cow Lane, which was further down the South End close to Fort Hill. Samuel G.

was only a block away from his Milk Street tenement. (Figure 12) Documents clearly show that he was renting this place in 1707, even though the questions of when he started and how long he stayed cannot be answered. He most likely continued some work at home and stored material and products in his shed and cellar. But the rented new shop was certainly the place where he would like to display the Blue Ball and to meet his customers.

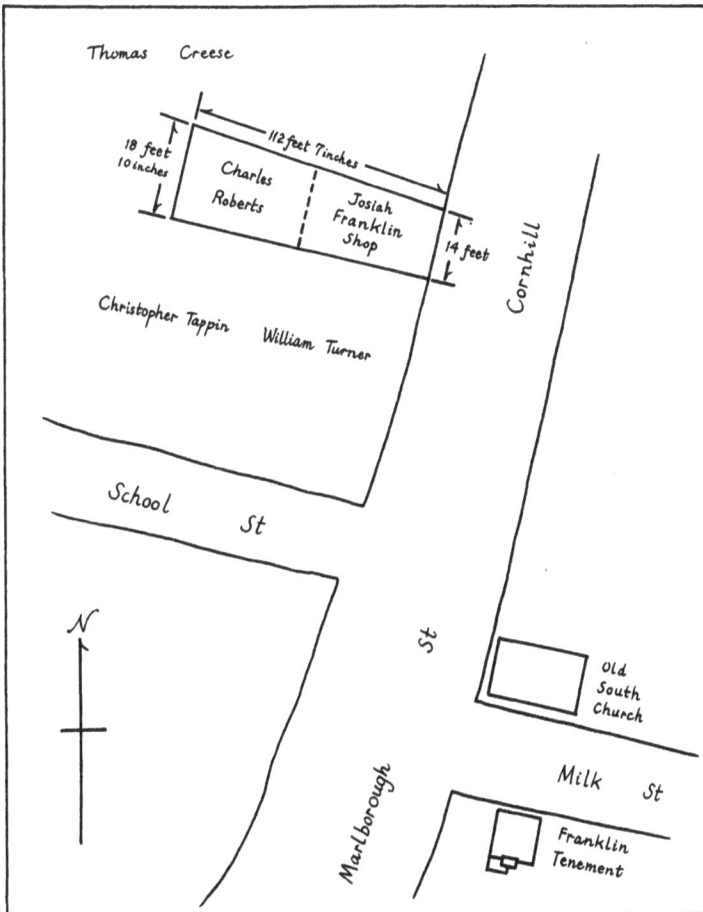

Figure 12. Franklin's tenement and shop, according to *Suffolk Deeds*, 23: 147.

Drake, *The History and Antiquities of Boston* (Boston, 1856), Appendix No. II, pp. 807, 818.

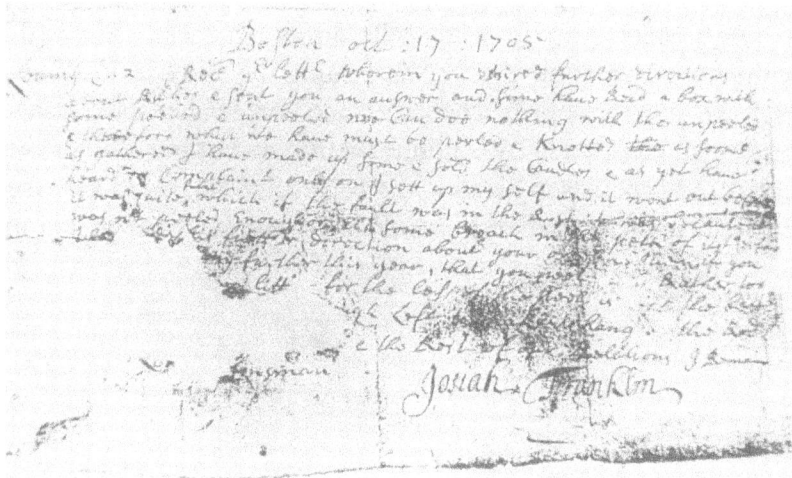

Figure 13.   Josiah Franklin's letter to [Peter Folger, Jr.] October 17, 1705.

Always mindful of finding new ways to promote the business, Franklin also tried to duplicate the English practice of making rush candles. In a letter to Peter Folger, Jr., of Nantucket on October 17, 1705, (Figure 13) he wrote: "Loving Cuz, I Recd your lette wherein you desired further directions about Rushes & sent you an answer and since have Recd a box with some peeled & unpeeled." He warned, however, that he could "doo nothing with the unpeeled & therefore what we have must be peeled & knotted as Soone as gathered." He made and sold some of the rush candles and noticed that one had burned out before it should. "If the fault was in the Rush," he pointed out, "it was because it was not peeled enough or else some breach in the peth." Knowing this to be a very gentle and methodical task, he urged Peter to peel the rush properly, neither too little or the rush would not burn, nor too much or the rush would not stand.[92] The collaboration apparently did not go well because several months later the disappointed Franklin wrote again: (Figure 14)

> Loving Cousin,
> I Received yours and I think I did send word that I received
> the Rushes but am not certain, but now I do and thank you
> and cousin John for your care and pains about them and am

---

[92]Josiah Franklin [to Peter Folger, Jr.], October 17, 1705, Nantucket Historical Association, see Appendix, Document 4.

freely willing to pay for them and they do pretty well generally. The smallest was well peeled, but rather too small the biggest which I think was of cousin Johns gathering, was not quite peeled enough. I think to send one of my family thats most used about them to gather some (for they will not peel here and I find it will not be worth the while to take care of them) to see them well ordered.[93]

Figure 14.   Josiah Franklin's letter to [Peter Folger, Jr.] July 14, 1706.

Compared with the ill-fated rush-candle enterprise, the Franklins' secret recipe of making a fine crown soap fared much better and for many years remained a source of family pride.[94] Unlike those simple

---

[93]Josiah Franklin [to Peter Folger, Jr.], July 14, 1706, see Appendix, Document 5. This Peter Folger was perhaps Peter Folger 3rd (1674-1707), whose father Eleazer Folger (1648-1716) was Abiah's older brother. "Cousin John" was John Folger (1659-1732), son of Peter Folger 1st, another older brother of Abiah. *New England Historical and Genealogical Register*, vol. 16, pp. 274, 272; *The Papers of Benjamin Franklin* (New Haven, Conn., 1959), vol. 1, pp. liii-lvi, lxxi. The family member Franklin said to send could be Elizabeth who was old enough, before the migration, to remember the traditional English practice of making rush candles back in Banbury.

[94]Based on family correspondence, scholars generally thought that John Franklin was the inventor. He left for Providence in Rhode Island about 1719 and moved back to

rushlights that were aimed at the lower end of market, fine soaps were a luxury designed for middle to upper-middle type of families. Ordinary colonials often took time in the spring to make their own soaps, a very slow process of many days. Yet, what they were able to accomplish was to make soft soaps, a sort of product suitable only for rough use.[95] Fine soaps, on the other hand, required quality material, good water, and all proper executions. On occasion, even some of the most experienced soap makers were bewildered, not knowing why their hard soap either failed to come or soon disintegrated. Substantial households could buy toilet soaps when cleanness and personal hygiene became a serious concern;[96]   Boston merchants helped to fulfill their need by importation. Castile soap, named for the place it was produced in Spain but later corrupted into "castle soap," appeared in newspaper advertising as early as the 1720s. Made of olive oil with a soda base, obtained from the ashes of sea-weeds, castile soap was best known for smoothness and plasticity, while its natural smell and marbled veins adding to the product's attractiveness and fame.[97]

Boston a decade later. An early advertisement of the crown soap appeared in Franklin's *Philadelphia Gazette* in November 1744, when the soap was already perfected. The new soap, therefore, may have been tested first in the 1720s and 1730s. It would be hard to imagine that the father had nothing to do with it during this period, even though his exact role is now impossible to determine. Carl Van Doren, ed., *The Letters of Benjamin Franklin & Jane Mecom* (Princeton, N.J., 1950), pp. 204, 206, 250.

[95]George Francis Dow described, "Most farmers' wives dreaded soapmaking. It was one of the hardest day's work of the year. Usually it was made a point to have the soapmaking precede the spring cleaning." *Every Day Life in the Massachusetts Bay Colony* (1935; New York, 1967), p. 97. See also Laurel Thatcher Ulrich, *A Midwife's Tale: The Life of Martha Ballard Based on Her Diary, 1785-1812* (Vintage Books ed.; New York, 1991), pp. 264, 309, 402n.

[96]After his courtship of Madam Kartharine Brattle Eyre Winthrop failed, Samuel Sewall comforted himself noting that "her Dress was not so clean as sometimes it had been." M. Halsey Thomas, ed., *The Diary of Samuel Sewall* (New York, 1973), vol. 2, p. 967.

[97]Campbell Mortif, *Chemistry Applied to the Manufacture of Soap and Candles* (Philadelphia, 1847), pp. 199, 201, 205; Philip Kurten, *The Art of Manufacturing of Soaps* (Philadelphia, 1854), pp. 183-84; Adolph Ott, *The Art of Manufacturing Soap and Candles* (Philadelphia, 1867), pp. 78-80; *Haney's Soap-Maker's Manual* (New York, 1869), pp. 10-11; William Lant Carpenter, *A Treatise on the Manufacture of Soap and Candles* (New York, 1885), pp. 192-193; Edgar George Thomssen, *Soap-Making Manual* (New York, 1922), pp. 79-81; and David T. A. Linday, *Soapmaking Past and Present* (1971; Nottingham, England, 1979), pp. 2-5.

A lack of these foreign advantages did not discourage the Franklins who utilized native resources to the best of their ability. According to the youngest child in the family, Jane Mecom's descriptions, two conditions must be met to make the soap. The first was quality material, including thirty pounds of "clean hard tallow" and fifteen pounds of "the purest bayberry wax of lively green color." Another key was that in addition to the common steps of making hot soap, the Franklins invented a new procedure called "separation," which meant to separate the soap from the lye used in the initial stages.[98]

The final product was excellent, most suitable for washing fine linens and other bright and delicate material.[99] In fact, this special soap became such a desirable commodity that even his runaway son Benjamin was quite willing to sell it in Philadelphia.[100]

🐲 "He was a pious & prudent Man"🐲

At the corner of Milk Street and forty feet across from Josiah Franklin's tenement stood a cedar-wood meeting house, the Third Church in Boston, an institution he was affiliated with for life. Commonly called the South Church or Old South (Figure 15) after

---

[98]Once a hot soap has come in consistence, continue to boil it hard in a copper and proceed to separation by sprinkling salt at the speed of one peck for every fifteen minutes. All the while, keep stirring from the bottom with a stick. Frequently check the stick and observe the drops from it. When the soap cools on the stick and its color becomes as clear as Madeira wine, the separation has taken place. Add no more salt, or the soap will be brittle. Continue to boil the copper fast but carefully keep it from boiling over until all the froth boiled in. Now the soap is separated and will lie on top of the salt lye. Clear away the fire with water, and put one pail of new lye into the copper to help the soap to rise. Leave the copper still until the next morning when the soap will be cold on top like a hard cake. Cut off the soap cake, wipe off any dirt or soil underneath, and throw away the salt lye left at the copper bottom. Wash and clean the copper, set it on fire again, and start the second round of separation by repeating the same steps as the day before. A slow fire shall suffice this time because of the weak lye found after the first boil. When all the froth is boiled in, add a pail of new lye and let the copper stand about half an hour. Now the soap can be poured out into a mold. To ensure quality, final cutting may begin only when the soap becomes lukewarm.

[99]Carl Van Doren, ed., *The Letters of Benjamin Franklin & Jane Mecom* (Princeton, N. J., 1950), pp. 125-34, 249-52. See also Claude-Anne Lopez and Eugenia W. Herbert, *The Private Franklin: The Man and His Family* (New York, 1975), pp. 40, 54, 108, 111, 244, 285.

[100]Paul Leicester Ford, *The Many-Sided Franklin* (New York, 1899), p. 11.

1817 when another church was built in Summer Street,[101] it was
founded by a group of twenty-nine people seceding from the First
Church in the aftermath of the controversy over the Half-Way
Covenant in the 1660s, and whose dissatisfaction was exacerbated by
the church's handling of the incoming minister John Davenport.[102] The

Figure 15. Old South
Church, building constructed
in 1729-30, viewed from
across Washington Street
after a recent renovation.
Photographed in spring 1998.

[101][Hamilton Andrews Hill], *The Early History of the Old South Church, Boston* ([Boston]
1891), vol. II, n. p. Everett W. Burdett, *History of the Old South Meeting-House in Boston*
(Boston, 1877), p. 11.

[102]The twenty-nine names were William Davis, Hezekiah Usher, John Hull, Edward
Raynsford, Peter Bracket, Jacob Eliot, Peter Oliver, Thomas Brattle, Edward Rawson,
Joshua Scottoe, Benjamin Gibbs, Thomas Savage, Joseph Rocke, Theodore Atkinson,
John Wing, Richard Trewsdale, Theophilus Frarye, Robert Walker, John Aldin,
Benjamin Thurston, William Salter, John Morsse, Josiah Belcher, Seth Perry, James
Pemberton, William Dawes, Joseph Davis, Thomas Thatcher, and Joseph Belknap.
Benjamin B. Wisner, *The History of the Old South Church in Boston in Four Sermons*
(Boston, 1830), pp. 8, 76-77. Hamilton Andrews Hill and George Frederick Bigelow,
comps., *An Historical Catalogue of the Old South Church (Third Church), 1669-1882*
(Boston, 1883), pp. 5-6. See also John Gorham Palfrey, *History of New England*
(Boston, 1865), vol. 2, pp. 486-92; vol. 3, pp. 81-85.

secession was not easily forgiven. The First Church, though invited, refused to assist the installation of the Reverend Thomas Thatcher, the first pastor of the new church, on February 16, 1669. For more than a decade, no ecclesiastical communication existed between the two churches. The strained relations began to thaw after Edward Randolph, special envoy appointed by the Lords of Trade, arrived in 1676. Within less than two years, as a result of his numerous reports to London, charging among others things, the ill-treatment Anglicans received in Massachusetts, the Lords of Trade decided that "the Lord Bishop of London bee directed to appoint forthwith some able Minister to goe and reside at Boston in New-England."[103] Alarmed, established Congregational churches felt a need to restore unity. The First and Third Churches began to correspond and soon resolved "to forgive and forget all past" and to live in "peace."[102]

When, in 1685, Franklin expressed his desire to "offer himself unto the covenant," meaning to apply for church membership,[103] he and fellow parishioners sensed that the political environment was becoming very precarious. Crown officers of the court of Charles II had already begun the *quo warranto* proceedings against the colony, which, pressed by royal commissioners and Randolph, now collector of the customs and secretary of the province, ultimately led to the abrogation of the

---

[103]Edward Randolph, *The Ill Treatment the Church of England Receives in Boston*, January 15, 1678/9, and the subsequent decision by the Lords of Trade, February 6, 1678/9, in Robert Noxon Toppan, ed., *Edward Randolph*, 7 vols., (Boston, 1898-99), vol. 3, pp. 34-36, 36-38. For documents on the issue of the Church of England in New England, see ibid., vol. 2, pp. 311-320; vol. 3, pp. 332-35; vol. 4, pp. 51-59, 65, 88-91, 150-53, and 305-310. On May 15, 1686, the Reverend Robert Ratcliffe, an Oxford graduate and first Anglican minister, arrived on board the royal frigate *Rose* along with Randolph. Chaplain." André Mayer, *King's Chapel The First Century, 1686-1787*, pp. 3-4.

[102] Benjamin B. Wisner, *The History of the Old South Church in Boston in Four Sermons* (Boston, 1830), p. 17.

[103]Each colonial Congregational church had a written covenant signed by original subscribers; "the church" meant its membership. People who subsequently wanted to join the church had to undergo a two-step process: 1) apply for admission to membership, termed "to offer themselves unto the covenant;" and 2) after examination, be accepted by the church, termed "to be covenanted and admitted to the communion." See [Worthington Chauncey Ford ed.], *Diary of Cotton Mather* (New York, [1911]), vol. 1, p. 549; vol. 2, pp. 13, 48. M. Halsey Thomas, ed., *The Diary of Samuel Sewall, 1674-1729* (New York, 1973), vol. I, p. 40, 40n.

original Massachusetts charter. A year later, the new King James II decided to tighten his grip on the American colonies and his appointee Sir Edmond Andros arrived with almost unlimited authority as governor-general of the newly-created Dominion of New England. As soon as his commission was read, this first non-Puritan governor took the ministers at the ceremony into his library and spoke to them about accommodating Anglican services in their meeting house.[104] The proposal was rejected. Early next spring, Andros sent Randolph for the keys of the South Meeting-House, the place of worship closest to his residence at the southwest corner of Wing's Lane (later Elm Street),[105] demanding that the doors be opened for the Anglicans.[106] Despite remonstrance, Andros and his retinue continue to do so whenever he was in town. Like the rest of the congregation, Franklin suffered the humiliation of waiting for the Anglicans to occupy their church first in the morning until their service could start in the afternoon, an agonizing experience that lasted for two years.[107]

Then, news came and Bostonians felt blessed that a Glorious

---

[104]The Lords of Trade suggested this policy as early as November 22, 1684, see their instructions *Concerning Coll. Kirk:–Quit-rents:–Churches &c.*, in Robert Noxon Toppan, ed., *Edward Randolph*, 7 vols., (Boston, 1898-99), vol. 3, p. 334.

[105]Annie Haven Thwing, *The Crooked & Narrow Streets of the Town of Boston, 1630-1822* (Boston, 1920), p. 81.

[106]An excuse for this action was the Puritans' continuous denial of Anglicans' request for land to set up their meeting place. Using his authority, Andros finally chose a public lot adjacent to a town cemetery in School Street, where King's Chapel was built. André Mayer, *King's Chapel The First Century, 1686-1787* ([Boston] 1976), pp. 6-7.

[107]Hamilton Andrews Hill, *History of the Old South Church* (Boston, 1890), vol. 1, p. 331. Benjamin B. Wisner, *The History of the Old South Church in Boston in Four Sermons* (Boston, 1830), pp. 17-18. André Mayer, *King's Chapel The First Century, 1686-1787* ([Boston] 1976), pp. 3-6. In his report to the Lords of Trade, Andros wrote,

> The Church of England being unprovided of a place for theyr publique woship, he did, by advice of the Councill, borrow the new meeting house in Boston, at such times as the same was unused, until they could provide otherwise; and accordingly on Sundays went in between eleven and twelve in the morning, and in the afternoone about fower; but understanding it gave offence, hastened the building of a Church, w'ch was effected at the charge of those of the Church of England, where the Chaplaine of the Souldiers performed divine service and preaching.

To the Right Hon'ble the Lords of the Committee for Trade and Plantations. The State of New England under the government of Sir Edmond Andros, 1690, in Charles M. Andrews, ed., *Narratives of the Insurrections, 1675-1690* (1915; New York, 1959), p. 230.

Revolution had ousted James II.[108] On the morning of April 18, 1689, about eight o'clock, the Franklins heard rumors that a certain portion of the town was armed: The South End thought the North was, and the North End the South. An hour later the drums were beat. A group of leading citizens, including Wait Still Winthrop, Samuel Shrimpton, and Peter Sergeant, quickly met upstairs in the Council Chamber of the Town House. About noon, a Declaration of Grievances was read from the balcony, in which Andros's authority was denounced "absolute and Arbitrary" and his Dominion government "Illigal." The Grievances further alleged that his political conspiracy against the colony was to turn the latter into "slavery" while his religious plot was "Popery." The accomplishment of the current plight of the colony "was hastened by the unweared Sollicitations and slanderous Accusations of a Man, for his Malice and Falsehood well known unto us all." There was no need to name the man, who was the widely hated Randolph.

No sooner had the cheers and shouts in front of the Town House subsided when Franklin saw dozens of armed townsmen passing his house to join those a thousand strong who seized the South Battery not far from the end of Milk Street.[109] Andros, who narrowly missed the barge for his rescue sent by the royal frigate *Rose* in the harbor, was trapped in the fort with a few dozen redcoats. Mr. Nathaniel Oliver and Mr. John Eyre, two gentlemen messengers representing the citizen Council and both being members of the South Church, came along with the crowd and handed him a suggestive but firm notice which read, "Sir . . . surprized with the Peoples sudden taking of Arms. . . . We judge it necessary you forthwith surrender and deliver up the Government and Fortification." It was signed by Wait Winthrop, Simon Bradstreet, William Stoughton and twelve others. Badly out-numbered and not altogether clear about what did happen in England, he surrendered late that afternoon. The bloodless revolution in Boston

---

[108]For contemporaries' accounts of the unfolding of events and their backgrounds, see Charles M. Andrews, ed., *Narratives of the Insurrections, 1675-1690* (1915; New York, 1959), pp. 165-297. A general introduction of this period is David S. Lovejoy, *The Glorious Revolution in America* (New York, 1972). Still valuable is John Gorham Palfrey, *History of New England* (Boston, 1865), vol. 3, chs. 7-9, 11-15.

[109]Originally named the Fort Street, Milk Street was a major pathway leading from the downtown area to the Fort and the South Battery. Annie Haven Thwing, *The Crooked & Narrow Streets of the Town of Boston, 1630-1822* (Boston, 1920), pp. 18, 148-50.

was over the next day when His Majesty's Ship *Rose* and the Castle were peacefully subdued.[110] The swift change of events effectively ended Sir Andros's government as well as his intrusion of the South Church. It is not known how Franklin, who owned a gun and two swords, reacted to this tumultuous time. Records only show that his marriage to Abiah Folger was solemnized by the Reverend Samuel Willard later that year on November 25, 1689.

"His powers of mind were of superior order," members of the South Church praised their minister, an eminent one in New England. A Harvard graduate in 1659, he was at first minister at Groton for twelve years until 1676 when he moved to Boston. His learning was extensive and solid; his piety was consistent, devoted, self-denying, and confiding. "His discourses," admirers believed, "were all elaborate, acute, and judicious." At the same time, he was such a humane person that "he knew how to be a son of thunder to the secure and hardened, and a son of consolation to the contrite and broken in spirit."[111] He succeeded Thatcher as the second pastor and was believed to have maintained more balanced views during the Salem witchcraft hysteria in 1692, even though three of the seven judges on the Court of Oyer and Terminer were from his church.[112] In order to instill Calvinist doctrines into the minds of his flock, he gave a series of lectures, in familiar language, on the Westminster Assembly's Shorter Catechism to the children of the congregation. Later revised and enlarged, those lectures were delivered to the public once a month on Tuesday afternoon, (Figure 16) beginning on January 31, 1687/8. "Heard with

[110]Robert Earle Moody, ed., *The Glorious Revolution in Massachusetts: Selected Documents, 1689-1692* (Boston, 1988), p. 53. Michael G. Hall et al., eds., *The Glorious Revolution in American* (Chapel Hill, N.C., 1964), pp. 37-53.

[111]Benjamin B. Wisner, *The History of the Old South Church in Boston in Four Sermons* (Boston, 1830), p. 13.

[112]They were Wait Still Winthrop, Samuel Sewall, and Peter Sergeant. The other judges were Deputy Governor William Stoughton, Bartholomew Gedney, Nathaniel Saltonstall, and John Richards. All seven men were Council members and named by Governor William Phips. Sewall later recanted. While the minister read his apologies, he was "standing up at the reading of it, and bowing when finished" in front of the whole congregation of the South Church on January 14, 1697. See M. Halsey Thomas, ed., *The Diary of Samuel Sewall* (New York, 1973), vol. 1, pp. xxvi, 366-67. Benjamin B. Wisner, *The History of the Old South Church in Boston in Four Sermons* (Boston, 1830), p. 13.

Figure 16.   Pulpit inside the Old South Church seen from where the Franklins used to sit.

a great relish by many of most knowing and judicious persons both from town and college," the lectures continued until his sudden death in 1707.[113] The two hundred and fifty lectures were then published in a folio volume of more than nine hundred pages, entitled *The Compleat Body of Divinity*. Josiah and James Franklin were among the 450 subscribers. The father, the only tallow chandler in this group, ordered two copies.[114]

For many years, the church insisted on the orthodoxy that admitting anyone to communion required "satisfactory evidence of regeneration." The process first involved a careful examination of the candidate's belief and character by the pastor and other church leaders. Only after the applicant passed this initial stage would he be asked to prepare a verbal or written statement of religious views and conversion

---

[113]Benjamin B. Wisner, *The History of the Old South Church in Boston in Four Sermons* (Boston, 1830), pp. 13-14.

[114]Later in 1736, he also subscribed another minister Thomas Prince's *History of New-England* for two copies; son John ordered one.

experience to be presented before the church. The candidate then retired and his candidacy was deliberated by members of the church. If admitted, he would be allowed to partake in the Lord's Supper and to take God's holy covenant. If not, he had to repeat and demonstrate his conversion experience before the church membership in the future until one day he could convince them. Strict adherence to this rule and other disciplines explained the modest size of the church. For the first two decades since the founding, the church membership hardly reached three hundred. It was after nine years of probation that Josiah Franklin and his wife Abiah Franklin were finally accepted as communicants on February 4, 1693. (Figure 17) By then the Third Church had no more than four hundred full members.[115]

The rigorous standard and stringent practice had their reward. In addition to the fulfillment of his spiritual life, this closely-knit community provided critical social contact from which Franklin could build a stable customer base and find valuable resources in an era of personal credit and financing. On January 25, 1711/2, he bought a house on the south-west side of Union and Hanover Streets from rich merchant Peter Sergeant and his wife Mehitable (Mehetable), both were members of the South Church. For this two-story wooden building, the selling price was £320 but Franklin had only £70 in cash.[116] He turned to another rich parishioner Simeon Stoddard, who loaned him £250, using the house as collateral.[117] According to Stoddard's memorandum, Franklin paid back the money in full on January 28, 1722/3 and the mortgage was cancelled.[118] It turned out, however, that he was able to do so only because he mortgaged the house for a second time to Hannah Clarke for £220 a day later on January 29.[119] The grantor was the wife of merchant William Clarke, who was a member of the South

---

[115][Hamilton Andrews Hill and George Frederick Bigelow, comps.], *An Historical Catalogue of the Old South Church (Third Church), 1669-1882* (Boston, 1883), pp. 15-18. Benjamin B. Wisner, *The History of the Old South Church in Boston in Four Sermons* (Boston, 1830), pp. 58-60.

[116]*Suffolk Deeds*, 26: 108, see Appendix, Document 10.

[117]Ibid., 109, see Appendix, Document 11.

[118]Ibid.

[119]Ibid., 36: 191, see Appendix, Document 16.

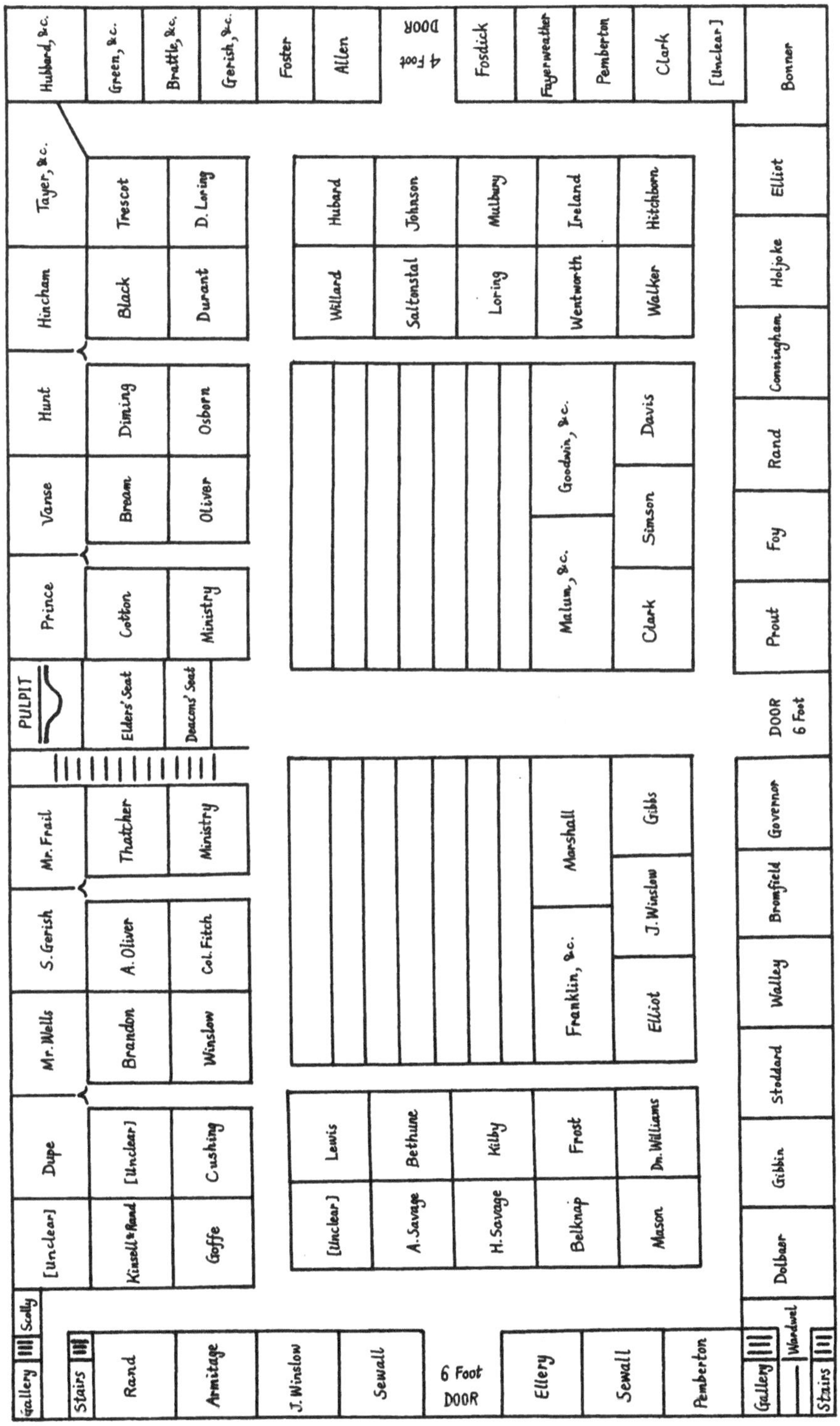

Figure 17. Seating arrangement of the Old South Church, based on a plan in Benjamin B. Wisner's *History of the Old South Church in Boston in Four Sermons* (1830).

Church. He died within a few years before 1726, and the widow married Josiah Willard, son of the Reverend Samuel Willard, who was a member of the same church and a provincial secretary for forty years. Considering the amount of money he borrowed against his income, age, and the value of the property, Franklin was fortunate to secure these two very substantial loans that might not be available otherwise. Knowing his finance as well as trusting his character, Franklin's guarantors were not disappointed. Hannah and Josiah Willard testified that their mortgagee paid back his loans in full on August 9, 1739.[120] He was eighty-two years old.

For a long time, parishioners of the South Church prided themselves on their neighborhood prayer meetings, suggested early by John Hull, the rich mint-master of the colony and one of the original twenty-nine founders. Samuel Sewall, who married Hull's only child Hannah Hull, wrote down numerous entries in his diary about those meetings, which began as early as 1676.[121] At first a monthly event and later a bi-weekly or weekly, the meetings took place at private houses in the early evening. A dozen or so parishioners rotated the site of their meeting from household to household, and different individuals were assigned to speak of a certain section from the Scriptures or of related topics. After Franklin became a church member, he participated. A particular night in early September 1708 seemed special to him, because Sewall recorded on the 8th that "I was mov'd last night at Mr. Josiah Franklin's at our Meeting, where I read the Eleventh Sermon on the Barren Fig-Tree. Tis the first time of Meeting at his House since he join'd."[122] Although both men were regular and sometimes the only participants, Sewall subsequently made ten additional references about Franklin and most of them were in the context of such private prayer meetings. He also revealed that Franklin was once a candidate for one of the two seats of deacon of the church, perhaps the one ambitious position he ever attempted. The diarist wrote on April 17, 1719, "South

---

[120]Ibid.

[121]See [Hamilton Andrews Hill and George Frederick Bigelow, comps.], *An Historical Catalogue of the Old South Church (Third Church), 1669-1882* (Boston, 1883), pp. vii-ix.

[122]M. Halsey Thomas, ed., *The Diary of Samuel Sewall, 1674-1729* (New York, 1973), vol. 1, p. 603.

Church Meeting p. m. Choose two Deacons; Mr. Barthol. Green, Mr. Dan¹ Henchman. Voters 41. Mr. Green had 37. Mr. Henchman 19. Mr. Franklin, 10."[123]

The defeat was hardly surprising because competitions for church offices depended on community status as much as personal piety. Bartholomew Green, son of the first colonial printer Samuel Green of Cambridge, was the able craftsman who brought out the first colonial newspaper *Boston News-Letters* in 1704. His second wife was Jane Toppan, a niece of Judge Sewall who, as their neighbor on Marlborough Street, mentioned the couple in more intimate terms and far more frequently than he did Franklin in his diaries. Daniel Henchman, an entrepreneurial bookseller, was no less ingenious in building his social connections. He married his daughter Lydia to his former apprentice and successful businessman Thomas Hancock, who later became one of the richest merchants in Boston.[124] Yet, the distinctions in wealth and society did not diminish these parishioners' mutual regard of individual integrity. When many years later John Draper, son-in-law of Bartholomew Green, received the news of Franklin's death on January 16, 1745, he immediately wrote in the newspaper he inherited from the father-in-law:

> Last night died Mr. *Josiah Franklin*, Tallow-Chandler and Soap-maker: By the Force of a steady *Temperance* he had made a Constitution, naturally none of the strongest, last with comfort to the Age of Eighty-Seven Years; and by an entire Dependence on his *Redeemer* and constant Course of the strictest *Piety* and *Virtue*, he was enable to die, as he liv'd, with Chearfulness and Peace, leaving a numerous Posterity the Honour of being descended from a Person, who thro' a long Life supported the Character of an *Honest Man*.[125]

Reminding all of how fragile life could be, mortality was a common and constant concern on the minds of many colonials, as

[123]Ibid., vol. 2, p. 921.

[124]This was the Hancock (1703-1764) who left the bulk of his huge fortune to his nephew, John Hancock (1736-1793), his junior partner. See Isaiah Thomas, *The History of Printing in America* (1810; New York, 1970), pp. 195-96, 201-02.

[125]*Boston News-Letter*, January 17, 1745.

countless funeral sermons, newspaper announcements, and private diary entries indicate. Still, Draper's praise stood out in that vast documentation because his publicized recognition was quite unusual at a time when most death reports in print were given to royal officials, provincial dignitaries, wealthy merchants, and gentlemen or gentlewomen. The publisher was certainly qualified to comment on Franklin for they used to be neighbors in Union Street. More than a simple announcement, his quick response and lengthy eulogy reflected his personal knowledge of the man and his sincere appreciation of Franklin's piety and honesty.

⚵ "The straitness of his Circumstances, keeping him close to his Trade"⚵

Reviewing the past from a modern era, some people tend to assume that Franklin's dull profession alone tells enough about his low social and economic standing, but this is not true. Colonial chandlers came from diverse backgrounds and fell into many different and sometimes overlapping social categories. Identifying Franklin as a tallow chandler without any context misses the varied dimensions of his livelihood. Simple and straightforward as though it may sound, the extent to which his life and activity connected with colonial society at large is not as plain as his trade title would suggest. In fact, what the trade title tallow chandler can tell us about his condition is far from easy to ascertain because he experienced many phases of the mixed social and economic strata in Boston.

Contrary to some modern assumption, candle and soapmaking in colonial time could be highly profitable and some makers did become prosperous and achieve social recognition and status. Tallow chandler Anthony Blount, for one, was a long-time active member of King's Chapel. One of the earliest subscribers to his church, he was twice elected warden in 1710 and 1711. He also became a founding member of the second Anglican institution Christ Church in the North End. Toward the end of his life in 1726, free of debt, he was able to will the entire estate to his wife Jane Blount after he bequeathed more than £400 in cash distributed among his mother, sister, nephews, and friends. Tallow chandler William Hough (Huff or Hufe) had a long established business in the North End, where he acquired sizable properties in Salutation Street.[126] Other tallow chandlers such as John Craister (Chraister) and Jeremiah Green acquired the title of gentlemen. Not every one of them made their entire fortune from candlemaking, of course; many were also merchants, retailers, shopkeepers, traders, and victuallers who found candle and soapmaking complementary to their businesses. Edward Barrett (Barratt), for example, was a shipwright as well as a candle-maker. Jeremiah Green perhaps made as much money from selling candles as from his frequent real estate dealings.

What hindered Franklin's rise to a similar economic and social status therefore was not his profession per se, but rather his constant struggle to raise and support his children, which can be better understood by comparison. Because of his small family size and limited obligation, another chandler Thomas Clark, though illiterate, had the luxury to buy expensive personal items for his own sake, such as a great coat of £8 and a short jacket with scarlet lining of as much as £25.[127] These two coats alone were more than half of the total value of Franklin's personal possessions listed in his inventory and the executor's account.[128] When Clark died in 1749, his inventory showed that the unsold candles, soap, and tallow amounted to more than £270, a very substantial sum that could easily purchase a sizable dwelling house. His impressive financial accounts clearly indicated how much a tallow chandler could make and accumulate at his prime, if only, that is, he did not father more than a dozen offspring.

Always struggling, however, the picture of Franklin's life was never so rosy. Having at least two rents to pay and many a mouth to feed, Franklin was always looking for tallow and other material all over town,[129] and his business could barely keep up with the basic needs of

---

[126]Samuel Sewall paid 9s. to Mr. Hough for his candles on October 13, 1688, *Account Book, 1688-92*, Massachusetts Historical Society. Hough's real estate holdings in Salutation Street later reached about 7,000 sq. feet. *Suffolk Deeds*, 18: 254, 21: 444, 23: 30, 23: 31, 27: 221, 35: 118, and 99: 202.

[127]See Appendix, Document 22.

[128]See Appendices, Documents 23, 24.

[129]Records show that Josiah Franklin was actively buying tallow in Boston. Townspeople seemed to know this well and when any cattle were slaughtered, they

his ever-growing family. His son vaguely described the father's trade as
a non-gainful employment and it remains unclear how many candles
and soaps he had to sell in order to make ends meet. We do know,
however, that he was unable to marry his eldest daughter Elizabeth
(1678-1759) until she was twenty-nine years old in 1707.[130] The
Franklins belonged to one of the lowest categories on the town's tax list.
In 1687, they paid 1s.10d. for taxes while Judge Samuel Sewall, their
rich neighbor in the same seventh precinct, paid 14s.10d.[131]

What marked Franklin's life was not the amount of wealth he
accumulated, but his ability to maintain a stable clientele, (Figure 18)
without which neither his business nor family could survive. At a time
when Boston's small population could not support more than a handful
of candle and soap-makers, he was one of the few who succeeded. The
turnover rate of candle and soap-makers, in fact, remained very high
throughout the eighteenth century. The first city directories show that
there were eight candle and soap-makers in 1789. Seven years later in
1796, only two of them remained.[132] Perhaps more than anything else,

---

would contact and sell him the tallow. See the testimony of Joseph Billings, January 3,
1715, Massachusetts State Archives, 8: 267. Arthur Bernon Tourtellot has pointed out
that Franklin might have been at first a tallow dealer and later moved into the business
of candle and soapmaking. *Benjamin Franklin* (Garden City, N.Y., 1977), pp. 52-53.
The earliest Puritans of the *Arbella* fleet brought forty cows with them and the town of
Boston maintained regular cow keepers from 1635 to 1741, see Benjamin W. Labaree,
Colonial Massachusetts: A History (Millwood, N.Y., 1979), p. 30; Robert Francis
Seybolt, *The Town Officials of Colonial Boston, 1634-1775* (Cambridge, Mass., 1939).

[130]If the age of his children's marriages could be any indicator, his eldest son Samuel
(1681-1720) married at twenty-four in 1705, and second daughter Hannah (1683-
1723) at twenty-seven (second marriage) in 1710, and the third daughter Ann(e)
(1687-1729) at twenty-five in 1712, *The Papers of Benjamin Franklin* (New Haven,
Conn., 1959), vol. 1, p. lvii.

[131]*Report of the Record Commissioners of the City of Boston* (Boston, 1876), vol. 1, p. 124.

[132]The eight craftsmen in 1789 were Victor Blair, William Frobisher, Joseph Hamlurey,
William Knight, Robert Laffan, John Lovering, John Nottleton, and Isaac White. In
1796 there were fifteen: James Baker, Joshua Davis, Gideon French, Benjamin Gault, --
Higginbotton, Jacob Flake, William Frobisher, John Kennedy, James Kirkwood, John
Lovering, Joseph Lovering, Joseph Lovering, Jr., Robert Lovering, John Nochols, and
Erasmus Pierce. See *City Directories of 1789* and *1796, Report of the Record Commissioners
of the City of Boston* (Boston, 1886), vol. 10, pp. 171-212, 217-302. Arthur Bernon
Tourtellot suggested that there were three candle-makers in Boston in the 1650s. Given
the low demand in the early years and using those 1790 figures as an example, the ratio

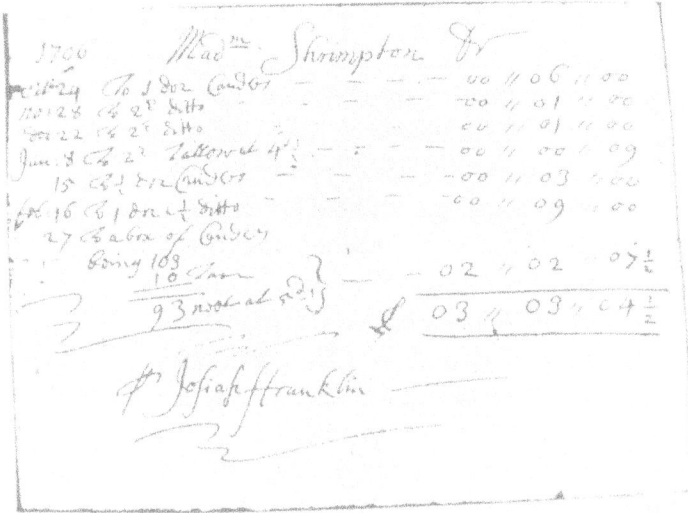

Figure 18.   Josiah Franklin's bill to Madam Shrimpton, 1706.

the remarkable longevity of Franklin's involvement in the business from the 1680s to the 1740s testified to his hard work and fortitude. He may well be the only tallow chandler in colonial history who survived for more than half a century in Boston.[133]

More intriguing than his financial status was the social standing of his profession, which cannot be determined solely by income or individual possessions. The simple skill and modest resources required in candlemaking enabled not only ordinary men but also housewives, servants, and sometimes slaves to conduct it. John Campbell, postmaster of Boston, put on this advertisement in 1706, "Four Negro Men Slaves, to be Sold, one about 18 years old, a second about 19, a third about 20, and a fourth about 30 years, that is a Tallow Chandler & a Soap-Boyler, and a good Work-man."[134] John Clark, tallow

---

of Boston population to candle-makers was about 2,250 to 1. This meant that around the turn of the seventeenth and eighteenth centuries when the population of Boston reached 10,000 to 12,000, the town could not support more than five to six full-time candle and soap-makers.

[133]Josiah Franklin never became a tradesman/merchant or a tradesman/land speculator as some candle-makers did. This could be the inflexible side of his long commitment to the trade, which, in addition to the large size of his family, may also explain his lack of more drastic financial advancement.

[134]*Boston News-Letter*, November 25, 1706

chandler, found no difficulty to allow his widowed daughters to continue the craft after his decease as he stated in his will that "my said wife Continue in the possesscon of the House I now live in with my two Daughters Ann Cook widow & Mary Jeffries widow in assisting them to Carry on the Business of a Tallow Chandler."[135]

No matter how inconspicuous his profession was, as an independent householder, Franklin's freedom and privilege gave him the opportunity to put up two advertisements in 1713. The first one read, "Three able Negro Men and three Negro Women to be sold by Messieurs Henry Dewick and William Astin, and to be seen at the House of Mr. *Josiah Franklin* at the Blue-Ball in Union Street near the Star Tavern, Boston."[136] It seems that Franklin was helping Dewick and Astin to sell their slaves at his house where the three men and three women might have stayed.[137] Dewick and Astin could be passers-by traders who needed a convenient local site to depose their slaves because a month later, a female slave was sold and the second advertisement appeared without their names, "Three Negro Men and two Women to be sold and seen at the House of Mr. *Josiah Franklin* at the Sign of the Blue Ball in Union-Street Boston."[138]

Similar activities of selling slaves, Indians, and white servants were commonplace in those days. Franklin's son John, for instance, owned a slave named Caesar, whom he willed to his wife Elizabeth in 1756.[139] Rich merchants such as Samuel Sewall,[140] Simeon Stoddard,[141] and

---

[135]*Suffolk Probate Records*, 25: 356.

[136]*Boston News-Letter*, June 22, 1713.

[137]According to Henry Louis Gates, in *Africana*, (New York: Basic *Civitas* Books, 1999, p. 287), slaves arrived in Boston in 1638 on the ship, *Desireé*. An anti-slavery sentiment grew strength around 1700, especially in the writings of Samuel Sewall, for example in his pamphlet, *The Selling of Joseph*. He also wrote in his diary for June 22, 1716, "I essay'd June 22, to prevent Indians and Negroes being Rated with Horses and Hogs; but could not prevail." For further reading on slavery in New England, see Lorenzo Johnston Greene, *The Negro in Colonial New England, 1620-1776* (Reprint Services Corp., 1942); William Dillon Pierson, *Black Yankees: The Development of an Afro-American Subculture in Eighteenth-Century New England* (University of Massachusetts Press, 1988).

[138]Ibid., August 3, 1713.

[139]*Suffolk Deeds*, 51: 93-98, see Appendix, Document 27.

[140]*Boston News-Letter*, April 9, 1716; December 17, 1716; May 25, 1719. *Boston Gazette*, May 30, 1726.

[141]Ibid., October 8, 1716.

William Harris[142] frequently advertised to sell slaves or Indians. Among Franklin's acquaintance and neighbors there lacked no profiteers who imported slaves and traded them on a regular basis. Jacob Royal, a merchant who lived at the opposite end of Union Street near the Town Dock, sold several dozen slaves in five years from 1726 to 1731 and Boston papers were full of his advertisement.[143] Female owners, such as widow Sarah Martin,[144] participated in this sort of transaction as well. Some did so perhaps in their effort to reduce household expenses, as in the case of Mrs. Margaret Franklin who was often behind her rental payment and therefore sold her slave servant after her merchant husband Henry Franklin Sr., passed away in 1713.[145] Modest business owners who were not slave traders but occasionally sold slaves could be found from almost every corner of the town, as the inundated newspaper announcements indicated: "Bartholomew Green, of the North End, had a slave boy of 14 for sale," "John Edwards, Gold Smith, of Cornhill Street, a slave woman," "Jarvis Bethell, Shoemaker, a slave girl of 18," "Thomas Downs, Tallow Chandler, a white servant," "Daniel Johonat, Stiller, a slave man," "Ambrose Vincent, Silk Dyer, a slave man and a slave woman of 19," "Capt. Thomas Smith, Dock Square, slave boy at 14," "Nathiel Breck, Union Street, a slave woman," "John Hobbs, Hanover Street, a slave man of 24," and "John Brocas, Union Street, a young slave man."[146] It is clear that Franklin's

---

[142]*Boston News-Letter*, March 30, 1713; May 10, 1714; June 21, 1714; August 9, 1714; October 8, 1716; July 29, 1717; May 19, 1718.

[143]*Boston Gazette*, May 30, 1726; August 15, 1726; September 5, 1726; April 3, 1727; May 15, 1727; September 9, 1728; October 7, 1728; August 18, 1729; November 10, 1729; August 16, 1731. *Boston News-Letter*, October 19, 1719.

[144]*Boston News-Letter*, October 13, 1718.

[145]Ibid., April 25, 1715. See also M. Halsey Thomas, ed., *The Diary of Samuel Sewall* (New York, 1973), vol. 2, 1035n. Henry and Margaret Franklin were not related to Josiah Franklin.

[146]For Bartholomew Green's advertisement see *Boston News-Letter*, September 27, 1707; John Edwards, *Boston News-Letter*, August 5, 1717; Jarvis Bethell, *Boston News-Letter*, October 15, 1716; Thomas Downs, *Boston News-Letter*, August 8, 1720; Daniel Johonat, *Boston News-Letter*, June 5, 1721; Ambrose Vincent, *Boston Gazette*, November 26, December 3, 1722; Capt. Thomas Smith, *Boston News-Letter*, December 23, 1717; Nathaniel Breck, *Boston News-Letter*, December 16, 1717; John Hobbs, *Boston News-Letter*, August 10, 1719; and John Brocas, *Boston Gazette*, January 27, 1729.

involvement resembled this last group of proprietors whose livelihood depended on their profession, but who did not hesitate to gain additional income by selling slaves.

As a free man and property owner, he was also qualified and elected to public offices. After Judge Sewall became one of the overseers of the poor, he recalled the time when he and other officials, including Franklin as a constable, went out to inspect Boston streets on a cold winter morning.[147] Throughout his life, Franklin was elected five times at the Town Meetings, three times as tithingman,[148] once as constable, and once as clerk of the market.[149] A low-ranking civilian enforcer of the law, his duties were not pleasant. He was obligated to watch the neighborhood with care, to spy on his neighbors if needed, and to help police a wide range of violations of the law, bad behavior, and lewd conduct, such as cheating in trade and at the market, domestic battery, idleness, drunkenness, profanity, and Sabbath breaking.[150] A thankless job, these offices often put him in the middle between the faceless laws and his familiar townspeople. Only a skilled tactician could fulfill his official duties while maintaining the respect and trust of his neighbors at the same time; the balance lay in prudence, fairness, and a willingness to act whenever needed. Never elected to any higher office, such as justice of the peace or selectman, he responded to every service call without complaint. His performance and experience in these regular duties gained him a small reputation feasible to discuss public issues with friends and even some "leading people" who, according to his son, "consulted him for his Opinion in Affaires of the Town or of the

<hr/>

[147]M. Halsey Thomas, ed., *The Diary of Samuel Sewall* (New York, 1973), vol. 1, p. 496.

[148]A tithingman (tythingman), originally meant one household header selected from every ten families, was a citizen watchdog whose principal task was to maintain peace in his neighborhood and to enforce order during the Sabbath. An English tradition, the office was first set up in Boston in 1676 and continued for a century until 1770.

[149]Once elected, some wealthier townsmen could afford to refuse the office by paying a fine. Either unable or unwilling or both, Franklin never chose to do that. See Robert Francis Seybolt, *The Town Officials of Colonial Boston, 1634-1775* (Cambridge, Mass., 1939), pp. 95, 100, 105, 139, 155.

[150]Arthur Bernon Tourtellot, *Benjamin Franklin* (Garden City, N.Y., 1977), pp. 67-70.

Church he belong'd to and show'd a good deal of Respect for his Judgment and Advice."[151]

While a basic fulfillment to both the family and the community was the best of what Franklin tried to provide,[152] he also explored a versatile personal life to the extent he could. His son recalled,

> He was ingenious, could drew prettily, was skill'd a little in Music and had a clear pleasing Voice, so that when he play'd Psalm Tunes on his Violin and sung withal as he sometimes did in an Evening after the Business of the Day was over, it was extreamly agreeable to hear.

A praecentor of the South Church himself, Samuel Sewall praised him twice for setting tunes well at their private meetings.[153] He also felt embarrassed when one day he set York tune, but the congregation went off into St. David's. "They did the same 3 weeks before," he wrote in the diary, "This is the 2^d sign. . . . This seems to me an intimation and call for me to resign the Praecentor's Place to a better Voice." Recognizing that after twenty-four years in that position, his musical gift was "being enfeeabled," he met the two ministers, his son Joseph Sewall and Mr. Thomas Prince and earnestly suggested that fellow church member John White or Franklin may replace him. "The Return of the Gallery where Mr. Franklin sat was a place very Convenient for it," he added.[154]

"On occasion," Benjamin Franklin indicated his father "was very handy in the Use of other Tradesmen's Tools." For example, his knowledge and skill in carpentry was such that fellow parishioner and next door neighbor Jeremiah Bumstead, a craftsman who had frequent

---

[151]*The Autobiography of Benjamin Franklin* (New haven, Conn., 1964), p. 55.

[152]For example, he and several others led a group of neighbors to lay out a drainage from Hanover Street down to Union Street and the Town Dock in 1726. *Report of the Record Commissioners of the City of Boston* (Boston, 1885), vol. 13, pp. 151-52, 161-62, 164.

[153]M. Halsey Thomas, ed., *The Diary of Samuel Sewall* (New York, 1973), vol. 2, pp. 876, 975.

[154]Ibid., 885-86. The position finally went to Mr. White, a Harvard graduate and clerk of the House of Deputies for twenty years, ibid., p. 985.

cash incomes, (Figure 19) entrusted son (Junior) to learn the trade from him and paid 40s. for his instructions. (Figure 20)[155] It seemed that he might even have entertained the idea to build his own frame house during the 1690s while the family was still renting. Considering the damage caused by fire to wooden buildings, the Massachusetts General Court passed a law in 1692, prohibiting future erections of wooden

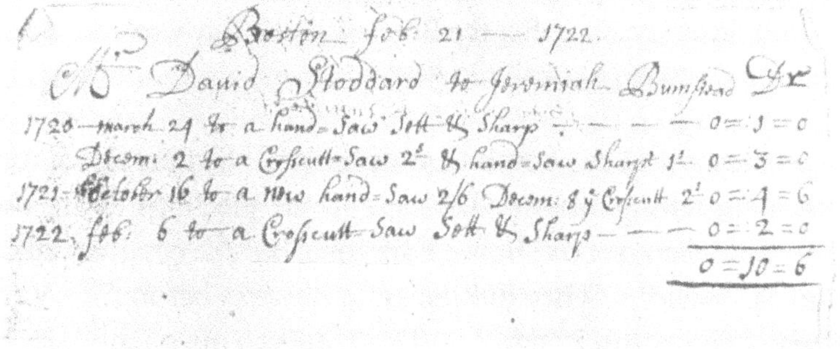

Boston feb: 21 — 1722

Mr David Stoddard to Jeremiah Bumstead Dr

1720 march 24 to a hand-Saw Sett & Sharp — 0=1=0

Decem: 2 to a Crosscutt-Saw 2s & hand-Saw Sharpt 1s 0=3=0

1721 October 16 to a new hand-Saw 2/6 Decem: 8 ye Crosscutt 2s 0=4=6

1722 feb: 6 to a Crosscutt Saw Sett & Sharp — 0=2=0

0=10=6

Figure 19. Jeremiah Bumstead's bill to David Stoddard, February 21, 1721/2.

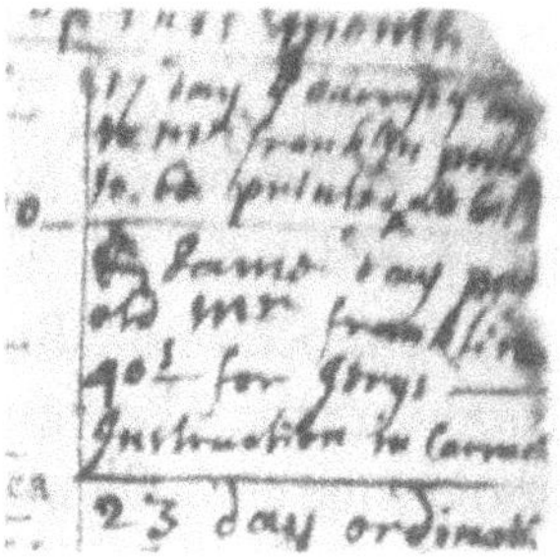

Figure 20. Jeremiah Bumstead's diary entry, in the margins of an almanac, "17. [May 1722] I carryd y^e cop[y] to Mr. Franklin, printer, to be printed, att 6:18. & same day paid old Mr. Franklin 40^s. for Jery's Instruction in Carpentry." (By permission of the American Antiquarian Society)

155S. F. Haven, ed., *Diary of Jeremiah Bumstead of Boston, 1722-1727, New England Historical and Genealogical Register*, vol. 15, p. 195. Compared with Franklin who could only collect a small amount of money after a period sometimes as long as more than a year, Bumstead's numerous entries in the diary about sizable cash gains seemed to be very impressive. The two knew each other well. A book lover, Bumstead often visited James Franklin's shop and requested print work; he was also an assessor who priced Benjamin Franklin the Elder's books after the latter's decease. His surviving bookplate can be found at the Massachusetts Historical Society. Jeremiah Bumstead Jr. was a witness to Josiah Franklin's last will of 1744, see Appendix, Document 21.

houses. It ordered that no dwelling house, shop, warehouse, barn, stable, or any structure of more than eight feet in length or breath and seven feet in height should be erected in Boston but of stone or brick and covered with slate or tile. Unless a special permit was granted by the justices or the selectmen of Boston, anyone who attempted to erect any frame or building contrary to the order shall be deemed a common nuisance. The owner shall demolish the structure, or he shall be committed to jail until he did; a fine of fifty pounds was lately added by the magistrates in 1699.[156]

Many townsfolk protested, including Franklin. He signed, along with two hundred and forty others, a petition against this new code. (Figure 21) They claimed that the law

> is found . . . to be greatly Injurious and Prejudiciall to the Generallity of the Inhabitants of this Towne, which consists, chiefly of Tradesmen who tho many of them have Lands and some Estate to befriend them, yet have not a Sufficiency to comply with the s^d Law and to carry on their Trades for the upholding their families.

"If wee have not speedy redress and relief herein," they warned, "many of us that are antient Inhabitants and Children of the first Planters and Setttlers of their place must either be forced for a subsistence and Livelyhood to Leave our Country, and that little our Fathers by their care, Industry and God's blessing thereon have left us, or else to become Tenenants to Foraigners that have come among us, & with their moneys, here purchased houseing and Lands, or att best to our Rich and Wealthy Neighbours."[157] Despite an initial denial of the magistrates, the new building code was never enforced as a result of the widespread protest, and the fear by many of becoming permanent tenants gradually subsided.[158]

---

[156]*New England Historical and Genealogical Register*, vol. 16, p. 84.

[157]Ibid., p. 85. See Appendix, Document 2.

[158]Thomas M. Babson, comp., *Special Statutes of the Commonwealth of Massachusetts Relating to the City of Boston, Passed Prior to January 1, 1893* (Boston, 1892), pp. 1-2, 2n. In the end, many signers of this petition built their timber houses, such as William Tilley, Joshua Gee, Isaiah Tay, and William Hough, *Report of the Record Commissioners of the City of Boston* (Boston, 1900), vol. 29, pp. 185, 190, 194, 197.

Figure 21.   Boston petition, November 18, 1696. Josiah Franklin's signature is in the third column right below the damaged area at the middle.

Franklin's participation clearly showed his desire to own a house. He only aborted the idea to build one when he purchased the Union Street estate sixteen years later. The event and his involvement also revealed his attitude and most likely association in town's affairs. Of those 241 signers, 152 had held town offices. Most of them were constables (71), tithingmen (70), and clerks of the market (32), the exact same offices that Franklin held in his career. Only eight of them had ever been selectmen and four representatives. In the meantime, both of those who held more important offices and those who held the lesser offices were few. The former category included two moderators, one tax collector, one town clerk, and one overseer of the poor; and the latter, water bailiffs (4), fence viewers (3), and bell-ringers (2).[159] The overwhelming majority of the subscribers were those who had inspirations for the future and yet had to strive hard to rise. These were precisely Franklin's peers.

🎐 "My Father now siding with my Brother"🎐

Moving to the Union Street in 1712 was a happy moment for the Franklins who, after renting for twenty-six years, finally had their own property. The wooden dwelling house, built in the 1640s or 50s, was as old as the tenement they left behind. Over the years the property changed hands at least eight times and was often occupied by tenants.[160]

---

[159]Counting the fact that one could hold more than one office at different times, the list of the signers showed that there were among them 71 constables, 70 tithingmen, 32 clerks of the market, 32 hogreeves, 21 surveyors of highways, 14 scavengers, 12 sealers of leather, 8 overseers of woodcords, 8 selectmen, 6 enginemen, 6 measurers of boards and timber, 6 surveyors of chimney, 5 assessors, 5 haywards and fence viewers, 4 lieutenants, 4 representatives, 4 watchmen, 4 water bailiffs, 3 fence viewers, 2 moderators, 2 captains, 2 inspectors of brick-making, 2 informers, 2 bell-ringers, 1 overseer of the poor, 1 rate commissioner, 1 town commissioner, 1 town gunner, 1 powder-keeper, 1 packer of flesh and fish, 1 measurer of grain, 1 prizer of grain, 1 sealer of weighs and measures, 1 tax collector, 1 town clerk, 1 engine master, and 1 overseer of the watch.

[160]According to *The Book of Possessions of the Inhabitants of Boston* in the City Clerk's Office, a manuscript volume compiled several years after the initial settlement but before a registry of deeds was established, shoemaker James Everill (Everell) owned "one house and houselott" that included "nearly the whole front on Hanover Street" in 1643. He divided the property into lots and sold them to several people, including William Tyng,

Furthermore, although the overall size was considerable, 96 feet on the north side, 38 on the east, 23 on the west, and 87 on the south, the property was divided into four lots. (Figure 22) The dwelling house sat on the largest one of 28 feet wide and 38 feet long at the south-west corner of Union and Hanover Streets.[161] If the new residence was

---

Francis Dowse, Even Thomas, William Correr, Robert Porter, John Stevenson, and William Hayward. One portion that he sold to Henry Maudesley about 1653 later became the Blue Ball property at the corner of Union and Hanover Streets. During the same year Maudesley fell into default and his property was taken by Edward Breck of Dorchester, who passed it to his son Robert Breck of Boston. Roger Seaward, seaman, bought the property in 1655 and held it for eighteen years before selling it to John Gill of Dorchester. Gill exchanged it to Lieutenant William Stoughton for a corn mill in Dorchester in 1673. Thirty years later in 1704, after Stoughton's death in 1701, the property was inherited by Mehitable (Minot) Cooper, who was the niece of the deceased, and her husband Thomas Cooper. Now owner of the Green Dragon and several adjacent properties, Thomas Cooper died at sea within a year and his widow Mehitable (Mehetable) married merchant Peter Sergeant, who sold the corner property to Josiah Franklin. For a printed version of *The Book of Possessions*, see Samuel G. Drake, *The History and Antiquities of Boston* (Boston, 1856), Appendix No. I, pp. 785-801; or *Report of the Record Commissioners of the City of Boston* (Boston, 1877), vol. 2, pp. 161-215. Also helpful is "Introduction: Estates and Sites," in Justin Winsor, ed., *The Memorial History of Boston* (Boston, 1882), vol. 2, pp. i-xlix. A summary description of the early transactions relating to the later Blue Ball property is Nathaniel B. Shurtleff, *A Topographical and Historical Description of Boston* (3rd ed.; Boston, 1891), pp. 626-37.

[161]Lacking detail, early descriptions of the property varied and, therefore, were insufficient to reconstruct its shape with precision. After tracing all different stages of transactions, Nathaniel B. Shurtleff believed that "it is sure there were as many as four [tenants] in all, as there were originally four tenements upon the estate." His conclusion was perhaps best supported by John Franklin's advertisement in 1753 when he was to sell his father's property:

> To be sold by public Vendue, on Tuesday the 21st of August next, Four Lots of Ground, with the Buildings thereon, fronting on Hanover and Union Streets, at the Blue Ball, viz. One Lot (No. 1) of Seventeen Feet Four Inches Front on Hanover Street, and twenty-five Feet deep. One ditto (No. 2.) Twenty-one and a half Feet Front on said Street, and Twenty-five and a half Feet deep. (No. 3.) Twenty-seven Feet Front on said Street, and Thirty Feet deep. (No. 4.) a Corner Lot, Twenty-eight Feet Front on Hanover Street, and Thirty-eight Feet Front on Union Street, very well situated for Tradesmen or Shopkeepers, being in the Heart of the Town, and the Buildings conveniently divided as above, having originally been different Tenements. The Title is indisputable; the sale to begin at four o'clock in the afternoon, on the Premises.

The phrase "with the Buildings thereon" was standard in language of most property transactions. But the description of "the buildings conveniently divided as above, having originally been different Tenements" was very critical in understanding this property.

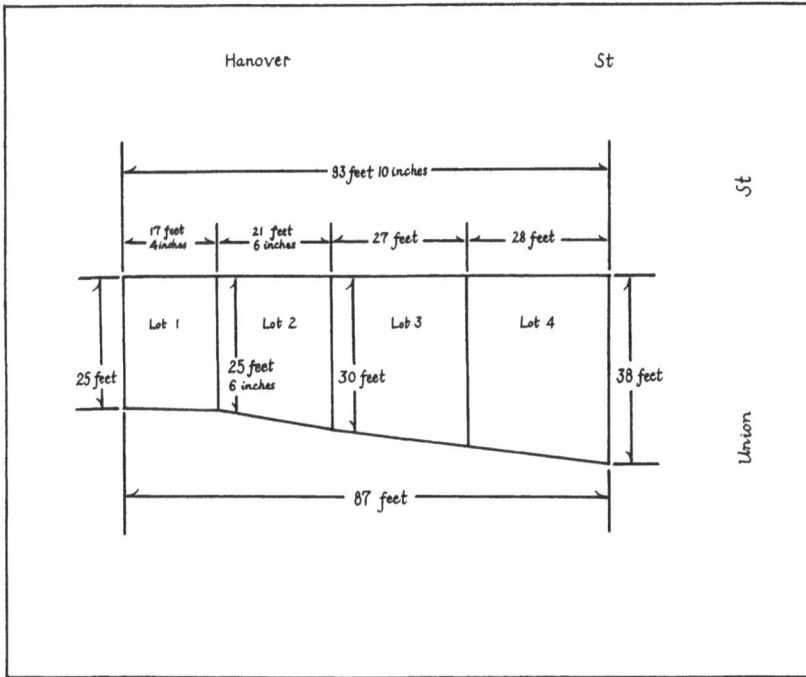

Figure 22.　Franklin property lines at the corner of Union and Hanover Streets, based on John Franklin's description in his advertisement on July 23, 1753.

bigger than the old, it was not by large measure. But the house had divided rooms, a real luxury to the owners.[162] In addition, four children

See Nathaniel B. Shurtleff, *A Topographical and Historical Description of Boston* (3rd ed.; Boston, 1891), pp. 630-33.

[162]Hardly has any description concerning the main house on the original corner lot survived; what can be certain is that, based on its value, the house was of a wooden structure. Meanwhile, Josiah Franklin's statement in his will about "the two rooms" of his house seemed to have confused some modern readers, because what he meant was "two stories," not two rooms in the modern sense of the word, which was generally called a chamber in his time. For the purpose of distinguishing a house of two floors, colonials would say the upper room(s) and the lower room(s). This expression was not registered in the *Oxford English Dictionary*, but had been clearly used by Franklin's contemporaries such as Samuel Sewall and Cotton Mather, see M. Halsey Thomas, ed., *The Diary of Samuel Sewall* (New York, 1973), vol. 2, p. 939, and [Worthington Chauncey Ford, ed.], *Diary of Cotton Mather* (New York, [1911]), vol. 2, p. 322. Few examples can be more unequivocal than a news report in the issue of March 4, 1723 for the *New-England Courant*, which stated that a great storm and high tide forced some townspeople to move to their upper rooms because the lower rooms were flooded. A

had left home: Samuel, the eldest son, married Elizabeth Tying in 1705; Elizabeth, the eldest daughter, married shipmaster Joseph Berry in 1707; Hannah, whose first husband was Joseph Eddy, married the

---

two-room house in Franklin's case, therefore, should be understood as a house of two floors, not of two chambers. Supposing it was a house of two rooms, there could only be a hall and a parlor on the same floor. If that were the case, Franklin would have been insane to pay the exorbitant sum of £320 to buy this sort of single-story house as he did.

Insofar as all the clues assembled and yet still without definitive proof, I suspect that the original house had a foundation of 26 feet by 36 feet, common to many houses of the same style of that period. It was a typical two-story colonial frame building similar to the Milk Street tenement for both were constructed during the 1640s and 50s. This building, however, had a central chimney and the entrance, doorway, and stairway were built in the middle of the house. Considering the size of the foundation, the house in fact had four rooms in modern terms: a hall and a parlor on the first floor and two partitioned rooms on the second.

Colonial houses centered around the hall where a hearth under the chimney was located. It could be the only room in small dwellings as the first floor of Franklin's Milk Street tenement exemplified. When space expanded, a parlor, used as a living room or bedroom or both, began to emerge as a portion of the house opposite the hall was separated. Because of this sequence of evolvement, the hall and the parlor were always opposite to each other on the same floor. Although the hall was sometimes called the great room and the parlor the best room, they were generally not considered as partitioned rooms in Franklin's time. A self-enclosed room, in the modern sense of the word, for either privacy or a particular purpose was more often referred to as a chamber, such as a butter-chamber, a bedchamber, and a maid chamber. Thus, it was altogether possible that an early colonial house did not have any room in the modern sense of the usage. But it was almost impossible for the same styled house to have more than one room but without a hall or a parlor. As a rule, a colonial dwelling house must have a hall and a parlor before it could have any chamber. Four large maps hung in his father's parlor, as the son specifically recalled, see P. M. Zall, ed., *Ben Franklin Laughing* (Berkeley, Calif., 1980,) p. 88.

Based on this understanding, the Union Street house can be envisioned to have four instead of two rooms simply because the central position of the chimney, the doorway, and the stairway naturally partitioned both floors into two sections. If the chimney was located at one end of the house, such as in the case of the Milk Street tenement, it would be difficult to heat the entire floor, a serious concern in a place where winter lasted long and hard. The central chimney solved the problem by dividing the house from the middle, and therefore could easily heat both halves of the house by using almost the same amount of fuel, not a small benefit in Boston where all firewood must be imported. Demanding some sophistication of carpentry to erect a framed house around the chimney, this arrangement was not a terribly more expensive construction than the Milk Street tenement, but made possible by a slightly different floor plan on a bigger foundation. A popular house structure since Franklin's time, it has remained so until today known as the colonial style throughout New England.

second time to Thomas Cole in 1710; and Josiah became a seaman who unfortunately died at sea. Another daughter, Anne, was engaged to William Harris of Ipswich and their wedding would take place in the summer of 1712. John, the eldest son with Abiah, was already twenty-one and the next son, Peter, twenty. Both would soon set up independently.

The house appeared to be as desirable and spacious as ever because the parents now had only three teenage children (Mary 18, James 15, and Sarah 13) and two toddlers (Benjamin 6 and Lydia 4) with them, a drastic change from what Benjamin used to "remember 13 sitting at one time at his Table."[163] Shortly after the family moved in, the youngest and last child Jane was born. Before the end of the decade, grandchildren were coming to visit, such as Samuel's and Elizabeth's daughter Elizabeth, Anne's and William Harris's daughters Anne and Grace, John's and Mary's son John, and Mary's and Robert Holmes's son William and daughter Abiah, a most pleasant scene described by Benjamin as "they maintained a large Family Comfortably; And brought up thirteen Children, And seven Grand Children Reput-ably."[164]

The Franklins purchased the property also for business considerations, not the least of which was the fact that some additional building structures existed on the rest of the lots. For the first time, Franklin could have his workplace, storage, and shop all at one site on his property. This time he once again hung the iron emblem at the corner of his new home, which became a permanent landmark of Boston for the next one hundred and fifty years—the Blue Ball.[165] Because of the enlarged space, the Franklins could take in not only occasional lodgers but also long-term tenants, both became an important new source of income. Such an improved living condition explained why Franklin was able to accommodate as many as eight people (Dewick, Astin, and their

---

[163]*The Autobiography of Benjamin Franklin* (New Haven, Conn., 1964), p. 51.

[164]Ibid., p. 56.

[165]This house was finally demolished in 1858 to widen Union Street for accommodation of increasing traffic. Ebenezer Whittier Stone (1801-1880), a veteran soldier and an adjunct general of the Massachusetts Militia kept the Blue Ball. The Massachusetts Charitable Mechanic Association preserved a chair made from the original timber of the building. Edward Everett, *The Mount Vernon Papers* (New York, 1860), p. 25. Samuel Adams Drake, *Old Landmarks and Historic Personages of Boston* (Boston, 1873), p. 147.

six slaves) on their property. It also explained why grocer Frederick Hamilton once stayed with the Franklins before he opened his shop in Union Street.[166] It also explained why Franklin was able to invite his brother Benjamin the Elder to come from London in 1715. He stayed not with his son Samuel, but with the Franklins for several years.[167] As Carl Van Doren has pointed out, these additional buildings made it possible for the Mecoms to occupy one of the tenements "close enough for Jane Mecom to be in regular attendance upon her aged parents."[168]

Finally, Franklin chose the property because he could see that conveniently located at the business and government section and at the southern tip of the North End, the new homestead was exposed to a dense population and busy traffic. On the Hanover Street side, the house was at a crossroads from downtown to the Winnisimmet (later Chelsea) Ferryway. On the Union Street side, it was two blocks from the conduit and three or four blocks from the Town Dock and the nearby Market Square, the heart of Boston commercial district. Right across the street on the northwest corner of Union and Hanover Streets stood the Star Tavern, a fine business establishment since the 1690s.[169] The singular attraction on Union Street, however, was a brick building—the famous Green Dragon. One of the most popular restaurants in town, it was frequented by the governor, his council, and other dignitaries and wealthy merchants.[170] Moving into the Union Street property literally meant becoming its neighbor, for the Green Dragon was located on the west side of Union Street just north of Hanover Street. It was only reasonable for Franklin to expect, therefore,

---

[166]*Boston Gazette*, March 8 and 15, 1731. Annie Haven Thwing, *The Crooked & Narrow Streets of the Town of Boston, 1630-1822* (Boston, 1920), p. 91.

[167]*The Papers of Benjamin Franklin* (New Haven, Conn., 1959), vol. 1, p. li. *The Autobiography of Benjamin Franklin* (New Haven, Conn., 1964), p. 48.

[168]Carl Van Doren, ed., *The Letters of Benjamin Franklin and Jane Mecom* (Princeton, N.J., 1950), pp. 39-40, 7.

[169]Annie Haven Thwing, *The Crooked & Narrow Streets of the Town of Boston, 1630-1822* (Boston, 1920), p. 89.

[170]M. Halsey Thomas, ed., *The Diary of Samuel Sewall* (New York, 1973), vol. 1, p. 544; vol. 2, pp. 620, 627, 632, 639, 640, 647, 670, 711, 716, 770, 775, 777, 791, 834, 854, 857, 891, 896, 930, 950, 970, 1004.

that the new homestead would make his Blue Ball more visible and more attractive to potential customers.

One problem left though, the steep purchasing price, which remained a heavy burden for the next thirty years. Despite the convenient new location and possible rental income, Franklin's savings rose slowly. Whereas he mustered £70 cash or an annual saving of £2 14s. from 1685 to the end of 1711 before buying the house, ten years later in 1722 he only managed to pay down the principal of his mortgage by £30 or an annual saving of £3, a slight increase of 6s. This small increase, however, must be viewed as an accomplishment at a time when many Bostonians suffered from contractions of employment opportunity and of silver-money supply. In fact, Gary B. Nash revealed that the value of the personal possessions of the lower and middle range Boston residents decreased during the several decades after Queen Anne's War. Franklin's ability to obtain a significant new property put him solidly in the category of the middle tier of Bostonians, whose total personal and real properties ranged from £87 to £300.[171] Franklin faired better than many townspeople because he was not a wage earner, who might have high skills but suffered from declining employment opportunities after the war, such as carpenters, shipbuilders, and mariners. An independent producer, his candles and soaps would continue to find a market as the population increased and as commerce with other colonies persisted.

His limited savings were significant also because of the large size of his family. The Franklins faced the incessant need to provide setting up funds for their four sons and dowries for their five daughters throughout this period. Perhaps no decade was more demanding than the 1710s when they simply had no time to relax: Hannah got married in 1710, John was twenty-one by 1711, and Anne's turn in 1712 and Peter's in 1713. The next daughter Mary's marriage would not come until 1716, or the next son James's maturity until 1718. Just as the parents could have taken a breath for 1714 and 1715, they sent Benjamin first to the Latin Grammar School and then to George Brownell's English School. The required twenty-four textbooks alone at the Grammar School would cost more than £6, and the first three

---

[171]Gary B. Nash, *The Urban Crucible: Social Change, Political Consciousness, and the Origins of the American Revolution* (Cambridge, Mass., 1979), pp. 19-20, 397-98, 399-400.

years of a seven-year curriculum wholly devoted to the Latin grammar had little to do with day-to-day survival.[172] After all, unless one seriously contemplated going to Harvard, there was no need to attend the Boston Latin which was a preparatory school for college. By the same token, those who did attend must be free from the worries of daily work and subsistence. Brownell's school taught elementary subjects, such as reading, writing, and cyphering. The Franklins' economic situation was such that they could not afford to finance one son through a complete schooling. Benjamin was withdrawn and went to work with the father. His strong dislike of the trade induced the father to compromise, which led to his apprenticeship to elder brother James.

It was a bold decision to support James to become the first printer in the family, whose setting up fund amounted to at least £100 as well as a trip to London in order to procure equipment. His sarcastic newspaper, the *New-England Courant* (1721-1726), brought much pleasure to his literary friends, but was an annoyance to the authorities who twice attempted to censure him. A resolution before the deputies on January 14, 1722 charged:

> The Committee appointed to Consider of the Paper called, *The New-England Courant*, published Monday the fourteenth Currant, are humbly of Opinion that the Tendency of the said Paper is to mock Religion, and bring it into Contempt, that the Holy Scriptures are therein profanely abused, that the Reverend and faithful Ministers of the Gospel are juriously Reflected on, His Majesty's Government affronted, and the Peace and good Order of his Majesty's subjects of this Province disturbed, by the said *Courant*; And for prevention of the like Offence for the future, the Committee humbly propose, *That James Franklin* the Printer and Publisher thereof, be strictly forbidden by this Court to Print or Publish the *New-England Courant*, or any Pamphlet or Paper of the like nature, except it be first supervised by the Secretary of this Province; And the Justices of His Majesty's Sessions of the Peace for the County of Suffolk, at their next Adjournment, be directed to take sufficient Bonds of the said *Franklin* for his good behaviour Twelve Months Time.[173]

---

[172]Pauline Holmes, *A Tercentenary History of the Bosotn Public Latin School, 1635-1935* (1935; Westport, Conn., 1970), pp. 258, 263.

[173]Quoted in Albert Matthews, *Bibliographical Notes on Boston Newspapers, 1704-1780*

Several months later on June 11, another line in the newspaper insinuated the government's connivance of piracy in their lukewarm effort to pursue a pirate vessel, which was a highly-sensitive political issue. The next day, a three-member committee again found that "the said Paragraph is high Affront to this government." James was therefore arrested and committed to jail for a month.[174] He felt ill within a week in the gloomy stone prison and professed that "he is Truly Sensible & Heartily Sorry for the offence he has Given to this Court." His poor health was certified by Dr. Zabdiel Boylston, a champion of inoculation whom he attacked in 1721. His petition was read and heard, but he was not released. After a bond security was given, he was allowed only to walk in the jail yard.[175]

The father was necessarily concerned in this ordeal. He set James up and signed two additional bonds with him to secure money and help the business. Both were dated on June 7, 1720, totaling almost £250. (Figure 23)[176] Obviously, thus heavily invested, the last thing he would like to see was more trouble. The friction between the two brothers, James and Benjamin, did little to please him. Early in July 1722, his own help "an Irish Man Servant, named William Tinsley" ran away and

---

(Cambridge, Mass., 1907), pp. 453-54. One of the three commissioners was Samuel Sewall who had a long and personal interest in the printing press. As early as October 12, 1681, the General Court granted him authority to manage the press in Boston, an office he held until 1684. He also owned a press and learned typesetting himself. When, in 1717/8, James Franklin went to London to become a printer, he carried a letter for Sewall (*Massachusetts Historical Society Collections*, ser. 6, vol. 2 [1888], p. 85). Sewall disliked the taste of the unofficial type of journalism from early on, and seemed to have no difficulty to vote for censure when he believed that "the Courant comes out very impudently." See M. Halsey Thomas, ed., *The Diary of Samuel Sewall* (New York, 1973), vol. 1, pp. 50, 56, 267, 460n; vol. 2, pp. 1004, 1107.

[174]Albert Matthews, *Bibliographical Notes on Boston Newspapers, 1704-1780* (Cambridge, Mass., 1907), p. 452.

[175]After he was finally released, James Franklin resumed his stinging satires and attacks. A sheriff, who had a warrant for his second arrest on January 28, 1723/4, failed to find him. Two weeks later on February 2, he was ordered to post a bond for his behavior, which was discharged on May 7, 1723. Clyde Augustus Duniway, *The Development of Freedom of the Press in Massachusetts* (1905; New York, 1969), pp. 97-103, 163-66.

[176]See Appendix, Document 20. After James died in 1735, his debt to the father was passed to his children who could not pay back until the grandfather's last days. Josiah canceled all the bonds in his will. See Appendix, Document 21.

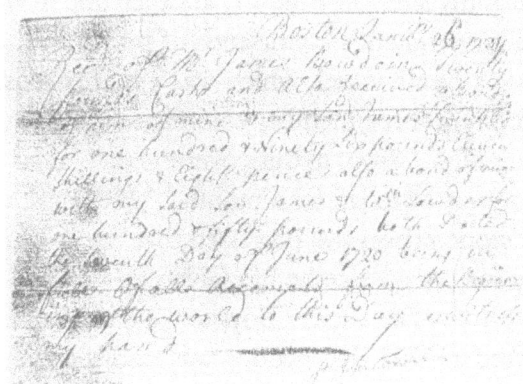

Figure 23.   Josiah Franklin's receipt from James Bowdoin, January 26, 1731/2.

he had to put up a forty-shilling reward for his apprehension and return.[177] Meanwhile Benjamin reached a secret agreement with James, which allowed him to take over as the publisher of the newspaper to evade government censure. Although he cleverly gave "our Rulers some Rubs in it," which James liked,[178] their unyielding temperament toward each other continued to undermine their relations. Neither James's "passionate blows" nor his apprentice's "saucy provocation" could ease the tension.[179] Finally, James knew that Benjamin was about to exploit his awkward situation and break from him, asserting "freedom" against the "tyrannical" master brother.[180] He went about town and talked other master printers into not hiring his brother. The father was also

---

[177]The full announcement read,

> Ran away from his Master Mr. *Josiah Franklin of Boston*, Tallow-Chandler, on the first of this instant July, an Irish Man Servant, named *William Tinsley*, about 20 Years of Age, of a middle Stature, Black Hair lately cut off, somewhat fresh-coloured Countenance, a large lower Lip, of a mean Aspect, large legs and heavy in his going; he had on when He went away, a Felt Hatt, a white knit Cap striped with red and blue, white Shirt and Neckcloth, a brown Colour'd Jacket almost new, a Frieze Coat of a dark colour, grey Yarn Stockings, leather Breeches trim'd with Black, and round to'd Shoes. Whosoever shall apprehend the said Runaway Servant, and him safely convey to his abovesaid Master, at the Blue Ball in Union Street Boston, shall have *Forty Shillings Reward*, and all necessary Charges paid.

New-England Courant, July 9 and July 23, 1722.

[178]*The Autobiography of Benjamin Franklin* (New Haven, Conn., 1964), p. 69.

[179]Ibid., pp. 68, 70.

[180] Ibid.

informed and this time, to Benjamin's surprise, he sided with James.[181]

Not deterred, Benjamin secretly left Boston in September 1723. He returned home from Philadelphia seven months later, this time "better dress's than ever . . . having a genteel new Suit from Head to foot, a Watch, and . . . Pockets lin'd with near Five Pounds Sterling in Silver."[182] His sudden reappearance delighted the family and everyone working in the brother's print office except James, who felt humiliated watching Benjamin's obnoxious display of his new found riches. How to be discreet and considerate Benjamin was still to learn. But he was also anxiously waiting for his father's response to a message by Pennsylvania governor William Keith, who suggested to set Benjamin up as a printer with the parents' permission and endorsement. The father seemed cautious and the son's wait long and disappointing.

"My Father receiv'd the Governor's Letter with some apparent Surprize; but said little of it to me for several Days," almost half a century later Benjamin vividly recalled that moment.

> When Capt. Homes returning, he show'd it to him, ask'd if he knew Keith, and what kind of a Man he was: Adding his Opinion that he must be of small Discretion, to think of setting a Boy up in Business who wanted yet 3 Years of being at Man's Estate. Homes said what he could in favor of the Project; but my Father was clear in the Impropriety of it; and at last gave a flat Denial to it.[183]

Little did he suspect that his father had untold difficulties.[184] With all the existing financial commitment, he signed another bond, (Figure 24) with his son-in-law James Davenport, to borrow £100 from a farmer Josiah Hobbs, which was due on May 1, 1724, the exact time when

---

[181]*The Autobiography of Benjamin Franklin* (New Haven, Conn., 1864), pp. 71, 70.

[182]Ibid., p. 81.

[183]Ibid., p. 82. Captain Holmes (Homes) was Robert Holmes (d. before 1743) who married Josiah's daughter Mary.

[184]The father did indicate to him that he had already advanced too much money for James, which was a sentence Franklin crossed out in his memoirs. But it is not known how much financial detail he revealed. *The Autobiography of Benjamin Franklin* (New Haven, Conn., 1964), p. 82n.

Benjamin returned.[185] The accumulated old and new debts made it impossible for him to spare any money for his youngest son, even if he had agreed with Sir William's scheme. Relieved to know Benjamin's

Figure 24. James Davenport's and Josiah Franklin's bond to Josiah Hobbs, April 1, 1725.

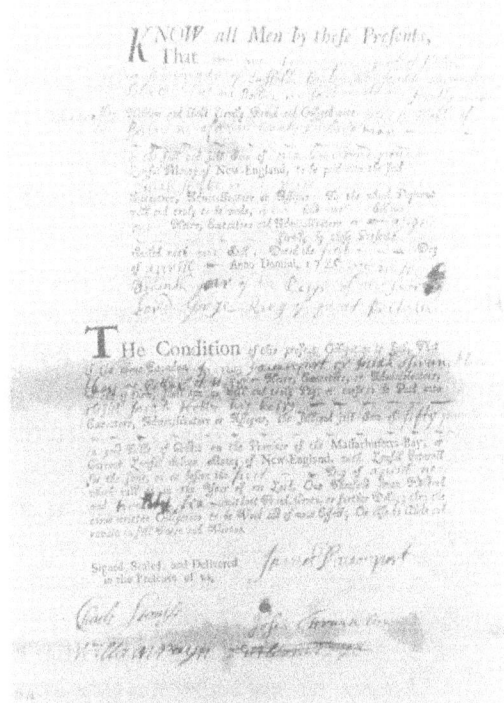

successful experience in the city of Brotherly Love, however, he advised "Industry" and "a prudent Parsimony" and gave his consent to the son's "Returning again to Philadelphia." "This was all I could obtain, except some small Gifts as Tokens of his and my Mother's Love, when I embark'd again for New York, now with their Approbation and their Blessing," he concluded.[186]

The father did hint that when the boy was close to twenty-one and when he saved enough to set himself up, he would help him out "with

---

[185]See Appendix, Document 17.

[186]*The Autobiography of Benjamin Franklin* (New Haven, Conn., 1964), p. 83.

the rest."[187] The promise did not materialize. His and Davenport's bond with Hobbs was extended for another year when, in 1725, both failed again to pay on schedule. Several years later the creditor still could not get his money back; his patience was exhausted. He brought the two debtors to court and the case was finally settled in 1729. (Figure 25)[188]

Figure 25.   Middlesex County warrant for Josiah Franklin's appearance at court, February 13, 1729.

About this time, Benjamin was planning his own enterprise in Philadelphia. His good business sense and personal credibility secured all the funds he needed from two friends, William Coleman and Robert Grace.[189] He soon became the sole proprietor of the *Pennsylvania Gazette* that would for the first time bear the imprint Benjamin

---

[187]Ibid.

[188]See Appendix, Document 19.

[189]*The Autobiography of Benjamin Franklin* (New Haven, Conn., 1964), pp. 122-23.

Franklin.

✿ "Their youngest Son, in filial Regard to their Memory" ✿

Perhaps out of a deep regret that Benjamin was the only child in the family whom he failed to support, Josiah Franklin wrote in his will that in addition to his one-ninth share of the estate, "I Give to my Son Benjamin in New Tenor to the Value of thirty Pounds Old Tenor."[190] This was a small but symbolic bequest because it was the only sizable amount of cash specified in the will by the aged father who would never forget his belated obligation. Little did he anticipate this time that the estate would become insolvent due to his and his widow's debt. His kindness toward Benjamin remained forever a tender memory and warm gesture.[191]

After mother Abiah Franklin passed away, Benjamin wrote to his brother John who now became the only surviving executor,

> when you have a little Leisure please to inform me how our Fathers Estate turns out as I hear every thing is now sold. Who bought the House, and what did it sell for? I feel some affection for that old fashioned Clock. It had I remember a sweet Bell; as it has been so long in the Family, I hope some of you have bought [it].[192]

John, at 63, had his infirmities and was having a hard time liquidating the parents' debt. An inventory taken on October 24, 1752 showed that except for the house and land valued at £253 6s. 8d., the rest of the estate amounted under £60, a clear sign of meagerness and insecurity.[193] A few months later, another account further indicated that

---

[190]See Appendix, Document 21.

[191]Carl Van Doren said that Benjamin gave both his share of the estate and the thirty-pound cash to his sister Jane. But he could not ascertain what she received. *The Letters of Benjamin Franklin and Jane Mecom* (Princeton, N.J., 1950), p. 40.

[192]To John Franklin, January 2, 1753, *The Papers of Benjamin Franklin* (New Haven, Conn., 1961), vol. 4, p. 409.

[193]See Appendix, Document 23.

all the personal property and receivables valued only £ 105 3s. 11d., not enough to redeem a debt of £166 7s. 11d.[194] John tried twice to dispose the property but because a court permission was yet to be obtained, his attempts received little response and stalled.[195] He then petitioned the Massachusetts Superior Court in 1754, which granted him the right to sell the house and land.[196] Brother-in-law William Holmes, who was the son of Mary Franklin and Robert Holmes and a goldsmith, finally bought the property at £188 13s. 4d. (Lawful Money),[197] which was about the estimated value of £250 (Old Tenor).[198]

As has been indicated, compared with the substantial inventory that a contemporary tallow chandler Thomas Clark left when he died, the Franklins' legacy appeared very meager indeed. Unlike their sons John and Benjamin, no portrait of them existed. One clue to their appearances came from Benjamin's pen, "He had an excellent Constitution of Body, was of middle Stature, but well set and very strong. . . . My Mother had likewise an excellent Constitution."[199] He also recalled that his parents rarely got sick.[200] They worked all their lives. To the Franklins, life was work. Hard work gave meaning and essence to life as well as its value and reward. Endless toils never bothered the father, who quoted Solomon's proverb, "*Seest thou a Man diligent in his Calling, he shall stand before Kings, he shall not stand before*

---

[194]See Appendices, Documents 24, 25.

[195]*Boston Gazette*, November 6, 1752 and July 23, 1753. See Nathaniel B. Shurtleff, *A Topographical and Historical Description of Boston* (3rd ed.; Boston, 1891), pp. 632-34, 636.

[196]See Appendix, Document 25.

[197]*Suffolk Deeds*, 85: 64, see Appendix, Document 26.

[198]For a useful introduction of the colonial exchange rate between Lawful Money (Sterling) and Old Tenor (Massachusetts currency) see John J. McCusker, *Money and Exchange in Europe and America, 1600-1775: A Handbook* (Chapel Hill, N.C., 1978), pp. 131-45.

[199]*The Autobiography of Benjamin Franklin* (New Haven, Conn., 1964), pp. 54, 56.

[200]Ibid., p. 56. Toward the end of the father's life or shortly after that in 1745, Franklin thanked his sister Jane "for your care of our father in his sickness," Carl Van Doren, ed., *The Letters of Benjamin Franklin and Jane Mecom* (Princeton, N.J., 1950), p. 40.

mean Men."[201] If the thought of standing before kings was hardly his initial drive to be industrious, the determination of never standing beneath idle bodies probably was. Trustworthy, consistent, and painstaking work dignified his life, who was convinced that nothing, not even well-intended work, "was useful which was not honest."[202]

No matter how difficult life was, he never stopped working or thought about moving to the countryside. He was a craftsman whose livelihood depended on his mechanical skills and talents. Quick in observation and adaptation, dexterous with tools, and mindful of innovations, he established an impressive record of sixty years of uninterrupted service in his trade. Essentially an urban person, he liked reading, music, and activity. Highly spiritual, he held a permanent pew in the South Church and devoted much time to attending sermons, public lectures, and private prayer meetings. Never an office seeker but an upright citizen, he dutifully fulfilled his civic obligations.

Like his brothers, Franklin was largely self-taught.[203] Often using a very fine pen, he had a good handwriting and kept clear accounts of business transactions.[204] A man of common sense, he understood the world from astute observations of real life and drew practical lessons from it. A book lover but never bookish, he was most congenial among his neighbors and peers, whose company and conversation he earnestly

---

[201]*The Autobiography of Benjamin Franklin* (New Haven, Conn., 1964), p.144.

[202]Ibid., p. 54.

[203]"Thomas was bred a Smith under his Father, but being ingenious, and encourag'd in Learning (as all his Brothers like wise were) by an Esquire Palmer, . . . he qualify'd himself for the business of Scrivener." *The Autobiography of Benjamin Franklin* (New Haven, Conn.,1964), pp. 47.

[204]Compared with other handwritings of the same period, his pen was always very sharp and indeed sharper than either that of Samuel Sewall's or of Benjamin Wadsworth's, which leads me to think that his pen could be something of his own device. It is truly amazing to see that some very tiny documents, such as Appendices, Documents 12 and 18, remained crisp and legible after almost three centuries. This fact might also have to do with the ink the Franklins used. Good dyers were extremely knowledgeable about how to make clear and durable colors. Benjamin Franklin the Elder expressed his pride about his ability to make black ink this way, "By this Repeit I make Ink in 1673. Which I saw was as black and Beautiful 30 years afterward as it was at first Writing." *Commonplace Book*, vol. 3, *Publications of the Colonial Society of Massachusetts* (1906), vol. 10, p. 206-07.

sought and sincerely enjoyed. He disdained a purely literary pursuit but took serious financial risks to save James and his printing enterprise. A no-nonsense speaker, he did not hesitate to poke fun at his relatives. Nevertheless, he was cautious enough to advise his son to "avoid lampooning and libelling to which he thought I had too much Inclination," according to Benjamin.[205] As a father, he indulged for a while in the thought of sending this son to the ministry. Even though the plan was aborted, he "exacted" Benjamin to attend church service as long as he was under his care.[206] Fearful of the prospect that the boy might become a seaman, he took him to visit various shops in order to engage him to a more suitable and less risky profession.[207] He also took time to read and comment on Benjamin's writings and it was "the Justice of his Remarks" that propelled the son to develop his means of self-improvement, a trademark of the Franklins.

They were a small and powerless clan in Boston, however.[208] Perhaps never dined once at the nearby Green Dragon or Star Tavern, the family was often struggling for survival and lacked the means and connections to move up in an increasingly stratified society.[209] Many

---

[205]*The Autobiography of Benjamin Franklin* (New Haven, Conn., 1964), p. 83.

[206]Ibid., p. 63.

[207]Ibid., p. 57.

[208]There were only a handful Franklins in colonial Boston and they did not seem to be related, such as William Franklin, blacksmith, who came to town in 1642, and Henry Franklin, merchant, who was affiliated with King's Chapel. Besides, Josiah Franklin came from Banbury, Oxfordshire, not a popular place of massive Puritan emigration, according to recent scholarship, such as David Hackett Fischer's *Albion's Seed: Four British Folkways in America* and Roger Thompson's *Mobility and Migration: East Anglian Founders of New England, 1629-1640* (Amherst, Mass., 1994). In fact, most of his relatives remained in the Church of England, which may explain the loneliness of the family situation in Boston. Next to power and wealth, having a sizable and well-connected network of clansmen was critical in a society that was much dependent on patronage and nepotism. The fact that Samuel Sewall could claim several dozen people his cousins was of definite social influence that no one could ignore. The Franklins had no such fortune. Even the second generation were not able to marry into powerful households. The family never became a significant clan in Boston.

[209]In February 1719/20, Samuel Sewall decided to help furnish his daughter Judith's house and made £50 available, which was equal to a half of the total value of Josiah Franklin's personal property. The purchasing list would dazzle the latter to a no small

years later, Benjamin returned to visit in 1753. Now a worldly man himself who succeeded in business and experienced in politics, he was delighted to see the pebbled streets and to hear the familiar accent.[210] He was even amused to recall the toy store where he bought the awfully expensive whistle and the unforgiving scolding that he subsequently received from siblings.[211] Both parents had passed away and the Blue Ball would be no longer the Franklins'[212] But something deeper never

---

degree for it included

> Curtains and Vallens for a Bed, with Counterpane, Head-Cloth and Tester, of good yellow watered worsted camlet (Send also of the same Camlet and Trimming, as may be enough to make Cushions for the Chamber Chairs), with Trimming, well made: and Bases, if it be the fashion; A good fine large Chintz Quilt well made; A true Looking Glass of Black Walnut Frame of the newest Fashion (if the Fashion be good), as good as can be bought for five or six pounds; A second Looking Glass as good as can be bought for four or Five Pounds, same kind of frame; A Duzzen of good black Walnut Chairs, fine Cane, with a Couch. A Duzzen of Cane Chairs of a different figure, and a great Chair, for a Chamber: all black Walnut; One Bell-mettal skillet of two Quarts: one ditto one Quart; One good large Warming Pan bottom and Cover fit for an Iron handle; Four pair of strong Iron Dogs with Brass heads, about five or six shillings a pair; A Brass Hearth for a Chamber, with Dogs, Shovel, Tongs and Fender of the newest Fashion (the Fire is to ly upon Iron); A strong Brass Mortar, that will hold about a Quart, with a Pestle; Two pair of large Brass sliding Candlesticks, about four shillings a pair; Two pair of large Brass Candlesticks, not sliding, of the newest Fashion, about five or six shillings a pair; Four Brass Snuffers, with stands; six samll strong Brass Chafing-dishes, about four Shillings a-piece; One Brass basting Ladle: One large Brass Ladle; One pair of Chamble Bellows with Brass Noses; One small Hair Broom sutable to the Bellows: one Duzzen of large hard-mettal Pewter Plates, new Fashion weighing about fourteen pounds; One Duzzen hard-mettal Pewter Porringers; Four Duzzen of small Glass Salt-cellars, of white glass, Smooth, not wrought, and without a foot; And if there be any Money over, send a piece of fine Cambrick, and a Ream of good Writing Paper; A Duzzen good Ivory-hafted Knives and Forks.

M. Halsey Thomas, ed., *The Diary of Samuel Sewall* (New York, 1973), vol. 2, p. 954n.

[210]To Samuel Mather, May 12, 1784, J. A. Leo Lemay, ed., *Benjamin Franklin: Writings* (New York, 1987), pp. 1092-93; To Rev. John Lathrop, May 31, 1788, ibid., pp. 1166-68.

[211]"The Whistle," November 10, 1779, in J. A. Leo Lemay, ed., *Benjamin Franklin: Writings* (New York, 1987), p. 932.

[212]William Holmes later sold the property for profit. By the end of the eighteenth century, the original wooden frame house had been torn down and a brick building constructed on the same foundation by a new owner, a descendant of blacksmith Jonathan Dakin. Several subsequent owners of this property continued to renovate the

disappeared: the Franklins' fortitude and resilience in life. Hoping that their work and example "shall not be wholly lost," he erected a new tombstone on their grave (Figure 26) and wrote these inscriptions:

Josiah Franklin
And Abiah his Wife
Lie here interred.
They lived lovingly together in Wedlock
Fifty-Five Years.
Without an Estate or any gainful Employment,
By constant labor and Industry,
With God's Blessing,
They maintained a large Family
Comfortably;
And brought up thirteen Children,
And seven Grand Children
Reputably.
From this Instance, Reader,
Be encouraged to Diligence in thy Calling,
And distrust not Providence.
He was a pious & prudent Man,
She was a discreet and virtuous Woman.
Their youngest Son,
in filial Regard to their Memory,
Place this Stone.
J.F. born 1655___Died 1744. Ætat 89
A.F. born 1667___Died 1752___85.[213]

reconstructed house until its demolition in 1858. Nathaniel B. Shurtleff, *A Topographical and Historical Description of Boston* (3rd ed.; Boston, 1891), p. 634. Some old photos showed a two-story brick building at the corner of Union and Hanover Streets, the Blue Ball hanging. But they were images of the house owned by those through the first half of the nineteenth century, not those of the one by the Franklins during the eighteenth. Later owners also gilded and repainted the Blue Ball, which can now be seen at the Bostonian Society.

[213]*The Autobiography of Benjamin Franklin* (New Haven, Conn., 1964), p. 56. Many years later, the tomb was repaired and a 20-foot granite obelisk erected with the following words on a bronze tablet, "The Original Inscription Having Been Nearly Obliterated/A Number of Citizens/Erected This Monument, As A Mark of Respect/For The/Illustrious Author, MDCCCXXVII." This gravesite in the Granary Burial Ground is the final resting place of Franklin's parents, his sixteen siblings, and Benjamin

Figure 26.   Tombstone of Josiah and Abiah Franklin and the Franklin family grave in Granary Burial Ground behind Boston Athenaeum.

Franklin the Elder.

APPENDICES

DOCUMENT 1

Samuel Sewall's Bill of Lading, April 2, 1689

(By Permission of the Massachusetts Historical Society)

Shipped by the grace of God in good Order and well conditioned by *Samuell Sewall of Boston In New England and for my own proper accott & risque* in and upon the good *Ship* called the *owners Adventure of Charlstown* whereof is Master under God for this present Voyage *Capt Nathaniell Cary* and now riding at anchor in the *Harbour of Charlstown* and by Gods grace bound for *Port Royall on Jamacia* to say *ten small barrells of tar, eight barrells of pickled pork, seven half barrells of pickled pork one box of Candles* - - - - - - being marked and numbred as in the Margin, and are to be delivered in the like good Order and well conditioned at the aforesaid Port of *Jamacia* (the dangers of the Seas only excepted) unto *Capt Nathaniell Cary* or to *his* Assigns, he or they paying fraight for the said Goods *five pounds*- with Primage and average accustomed. In witness whereof the Master of Purser of the said *Ship* hath affirmed to three Bills of Lading, all of this tenour and date, the one of which three Bills being accomplished, the other two to stand void. And so God send the good *Ship* to her desired Port in Safety, Amen.

Dated in *Boston in N: England 2 Aprill 89*

n° SS
1 to 26

primage
p^d

  Contents unknown   ℘ *Nath. Cary*

DOCUMENT 2

Boston Petition, November 18, 1696[1]

(By Permission of the Massachusetts Archives)

To the R^r Honble William Stoughton Esqr Liev^t Governo^r and Comander in Chiefe of his Maj^{ties} Province of the Massachusetts Bay in New England with the honrd Council and Representatives thereof now assembled in Gen^{ll} Court held at Boston by Adjournm^t November 18th 1696.

The Petition of us the Subscribers being Sundry of the Inhabitants of the Towne of Boston.

Humbly Sheweth,

That the Law relateing to building, with Bricks in Boston is found by continuall experience to be greatly Injurious and Prejudiciall to the Generallity of the Inhabitants of this Towne, which consists, chiefly of Tradesmen who tho' many of them have Lands and some Estate to befriend them, yet have not a Sufficiency to comply with the s^d Law and to carry on their Trades for the upholding their familyes; And there being much Land lying waste in This Towne, which if build upon would not only be advantageous to particular persons butt to the Publick also, by the Increase of the Towne.

That it is evidently and apparently seen by all Observing persons among us, that by this present long continued and Wasting warr, and through the Scarsity and dearness of Provisions wee are very greatly Impoverished and distressed soe that many of us know not how much Longer to supply the wants of our Poore familys and to comply with the demands of the Publick; And if wee have not speedy redress and relief herein many of us that are ancient Inhabitants and Children of the first Planters and Settlers of this place must either be forced for a Subsistence and Livelyhood to Leave our Country, and that little our Fathers by their care, Industry and God's blessing thereon have left us, or else to become Tenenants to Foraigners that have come among us, & with their moneys, here purchased houseing and Lands, Or att best to our Rich and Wealthy Neighbours, who are sometimes telling us, That if wee

---

[1]Massachusetts Archives, v. 113:139-40. A transcription was printed in the *New-England Historical Genealogical Register*, vol. 16, pp. 84-87. To understand the background of this petition, see "An Act for Building with Stone or Brick in the Town of Boston, and Preventing Fire," in Thomas M. Babson, comp., *Special Statutes of the Commonwealth of Massachusetts Relating to the City of Boston, Passed Prior to January 1, 1893* (Boston, 1892), pp. 1-2.

cannott comply with the Law wee must sell our Lands, which is a very hard and unreasonable thing, (Seeming much like to the Israelites Egiptian Bondage in makeing Bricks without Straw) seeing lime, slates or tiles are not to be purchased had wee Estates to comply with the Law.

Your Petition[rs] Therefore doe humbly Entreate That this high and hon[rble] Court will take the premises into Consideracon soe as that the s[d] Law relateing to Brick buildings in Boston may be repealed and utterly made Null and void in all respects.

And the Petition[rs] as in duly bond Shall ever pray.

Silvanus Davis	Jn°. Ricks	Sam. Bill
Samuel Bridge	Joseph Riall	Joseph Billings
Richard Keats	Joseph Adems	Tho. Phillips
Joseph Bisco	Tho. Stevens	Sam[l]: Pearce
Jonath. Evens	Jn°. Arnold	Tho. Roper
Jn° Walley	Benj[a]: Gallop	Hennery Cole
Jn° Combs	Josh. Hewes	Joseph Holmes
Tho: Stanbery	Peter Barber	Sam[l]. Flack
Joseph Gallup	Jono[th]: Berny	Joseph Vickers
Ebinezer Clore	Florence Mecarta	Edw[d] Keets
Tho: Cobb	Timo: Nash	Ebinezer Lowell
Richard Cobb	W[m] Thwinge	Tho. Powell
Wm: Porter	Jn° Parker	Wm. Gibbon
Tho: Lesenbee	Wm Holwell Jun[r]	Jn° Balston
Wm. Wheler	Anthony Greenhill	Tho. Baker
Tho: Harris	Obadiah Emons	Robert Sanderson
Wm: Holway	Rignel Odell	Edw[d] Taylor
Nich°: Sparrey	Eliezer Star	Josh. Lane
Richard Preist	Edward Durant	Danill Allin
Jn°: Pell	ho Oaks Jun[r]	Jn°. Marion Jun[r].
John Atwood	Return Wait	Rolan Story
Tho. Barnott	Jn°. Taller	Henry Rite
John Bennit	Joseph Tolman	Henery Mills
Peter Wear	Jn°. Farnum	Wm. Mamford
Samuel Marshal	Jacob Meline	Sam[l]. Bickner
Nath. Holmes	Wm: Robee	Tho. Wheler
Seth Perry	David Copp	Wm. Barage
Parcefell Clark	John Goodwin	James Fludd
Mathias Smith	Mich: Willis	Richard Flud
Sam[l]. Marion	Jn°. Clow	Joshua Gee
Jn°. Morton	Wm. Grigs	John Marion
Ralfe Ransford	Tho. Cushing	John Goodwin
Joseph Wheler	Ben Emons	Joseph Belknap
Wm. Tedman	Jn°. Winchomb	Bar. Arnald
Jn°. Cole	Richard White	Mic[h]. Shaller

Barth. Green
Jn°. Allen
James Harris
George Clark
Richard Paine
Jn°. Ranger
Wm. Clow
Jn°. Langdon
Saml. Grice
Jn°. Dinsdell
Nath. Goodwin
Arther Hael
Gamaliel Rogers
Nath: Baker
Elez^r Darbee
Ambros Dawes
Saml. Clowe
Joseph Jackson
Joseph Pearss
Tho: Savage
Jn°. Eustis
Jabesh Negus
Henry Ems
Wm. Gill
Jabes Salter
Arther Smith
Tho: Kelen
Dauid Norton
Newcom Blake
Joseph May
Phillip Finnee
Jn°. Jenkins
Andrew Cuningham
Jn°. Kneeland
Thomas Child
Andrew Mariner
Barth. Sutton
Saml. Earle
Thomas Gould
Wm. Frothingham
Peter Butler
Theophilus Frary
Joseph Elliot
Elisha Odlin
John Mason

Daniel Morey
Danill Phipenny
Tho. Peck sen^r.
Joseph Hill
Jn°. Claverly
Peter Warrin
Saml Gray
John Cutler
Jn° Fosdike
Josiah Franklin
Sarill Simson
Jn°. Berree
James Webster
Tho. Peck Junr.
James Andros
Thomas Walker
Thomas Lincoln
Robt. Earle
Ben Backworth
Mathu Delver
Saml Bridg Jun.
Saml. Weaver
Richard Hubbert
Sim°. Masinger
Jn°. Roberts
Ben. Fitch
Edwd. Bartlit
Wm. Wheler
Joseph Rodgers
James Thornbay
Joseph Lowell
Jn°. Clowe
Joseph Lowell Junr
Dauid London
Wm Tilley
James Labloon
Tho. Paine
Isaac Marion
Benja. Snelling
Benja. Bream
Obediah Read
David Farnam
Saml. Greenwood
Thomas Downes
Thomas Oakes
James Barns

Stephen Minott
W^m. Obbirson
Jn°. Humpherys
Richard Gridly
Josh. Cornish
Richard Richeson
John Tolman
Richd. Price
Henry Briteman
Richard West
Silenc Allen
Jn°: Nicholson
Robt. Smith
Richd. Partman
Wm: Hawkins
Ambrose Hanwe^{ll}.
Francis Moss
Ben. Holway
W^m. Meed
David Addams
Wm. Enecott
Richard Font
John Nichols
Raffe Carter
Jn°. Perrish
Wm. Hough
W^m. Tarnner
Sam^{ll}. Jacklin
Ebinezer Hayden
Edwd. Oakes
Thomas Baker
Richard Whitridge
Robert Seers
Jn°. Goffe
Jn°. Parram
Tho. Verny
Saml. Gardner
Nath. Alden
Rich. Way
Stephen French
Jn°. Child
Wm. Werden
Jn° Ball
Edwd. Ashley
Isaiah Tay

Wee the Selectmen of, (and For and in behalf of the Inhabitants of) the Towne of Boston doe humbly request and Entreate the favour of this high and hon^ble Court to Grant the above petition, Or if it may not seeme meet by yor hon^rs soe to doe That then you will please to grant That the Towne may have full power, authority & Free Liberty to choose such persons as they shall see meet from time to time to approve and allow or disallow of the place or places where Woodden Buildings in this Towne shall or may be, or may not be Errected & Sett up. And that all such persons as have hitherto Transgressed the Law, relateing to Brick buildings may not be lyable to Incurr the penalty thereof.

Pr Order of the Selectmen

Dated Boston June 11, 1697   Wm Griggs Town Cler:

Read the 11 June 1697

Read a Second time, 16th, & debated with the Report of the Committee thereon And Report Negatived.

DOCUMENT 3

Josiah Franklin's Bill to Boston for Candles to the Alms House, 1702/3

(By Permission of the American Philosophical Society)

1702/3   The Town of Boston Dr for Candles delivered to the Alms House

feb	26	To 2^E Candles	--	--	--	--	--	--	00-01-04
	11	To 4^E ditto	--	--	--	--	--	--	00-02-08
ap:		To 3^E ditto	--	--	--	--	--	--	00-02-00
June	21	To 3^E ditto	--	--	--	--	--	--	00-02-00
July	24	To 3^E ditto	--	--	--	--	--	--	00-02-00
aug	30	To 3^E ditto	--	--	--	--	--	--	00-02-00
oct	12	To 3^E ditto	--	--	--	--	--	--	00-01-10½
	27	To 3^E ditto	--	--	--	--	--	--	00-01-10½

$$00=15=09$$

Allowed [unclear] 1703   ℗   Josiah ffranklin

DOCUMENT 4

Josiah Franklin's Letter to [Peter Folger, Jr.],[1] October 17, 1705

(By Permission of the Nantucket Historical Association)

Boston Oct: 17: 1705

Loving Cuz   I Rec^d y^r lett^r wherein you desired further directions
about Rushes & sent you an answer and since have Rec^d a box with some
peeled & unpealed we Can doe nothing with the unpealed & therefore what
we have must be peeled & Knotted the as soone as gathered I have made
up some & sold the Candles & as yet have heard no Complaint only on[e] I
sett up my self and it went out before it was quite ˄^{burnt} which if the
fault was in the Rush it was because it was not peeled enough or Else
some breach in the peth of its [unclear]
    [torn]   further direction about your ordering them, if you
              any further this year, that you peel .. Rather too
        [t]oo litt[le] for the less of the peel is [unclear] the bett^r
                  [en]ough Left to make it hang .. the Rod
                  & the Rest of our Relations I remain
                  Lov[ing] [K]insman.

                              Josiah Franklin

---

[1]For references to the Folger's family see J. W. Jordan, "Franklin as a Genealogist,"
Pennsylvania Magazine of History and Biography, vol. 23 (1899), pp. 17-21; and the *New England Historical and Genealogical Register*, vol. 16, pp. 269-78.

DOCUMENT 5

Josiah Franklin's Letter to [Peter Folger, Jr.], July 14, 1706[1]

(By Permission of the Historical Society of Pennsylvania)

Boston July about the 14th, 1706

Loving Cuz   I Rec[d]. y[rs] & I think I did send word that I Rec[d] the
Rushes but am not Certain but now I doe & thank you & Cuz John for your
Care & paines about them & am freely willing to pay for them & they doe
pretty well generally, the smallest was well peeled but Rather too
small, the biggest which i think was of Cuz Johns gathering was not
quite peeled Enough-- I think to send on[e] of my family thats most
used about them to gather some for they will not peell here & I find it
will not be worth the while to take Care of them to see them well
ordered -- I have not Rec[d] Scollyes yett, & question whether Ever I
shall but I hear my patience hath som operation on his Ingenuity for I
hear by the by that he is Resolved if possible. I shall be paid but I
Could be glad to see it for its very low with him ___, according to the
desire I have housed 30-00-00 of your mony as I suppose Capt.
N[athaniel] G[ardner] Informed you as to the book of athiesm take it in
part of pay for the Rushes___
I Rec[d] the Verball lett[r]. by Br. Swain & I find you are in dang[r] of
Cooling for first you send me a pretty _____ letter & then they
grew __er & now it is dwindled up to word of mouth Therefore Lett us both
Endeavor to Reform & by ~~tht~~ that meanes I hope that we may doe somewhat
thus Iniquity may grow Cold & the Love of many may Abound which is all at
present from the Kinsman ----- Josiah ffranklin

my letter.[s] Longer then yours notwithstanding I had forgott till now to
Remember Love &c &c &c

---

[1] A transcribed and embellished copy is in the Folgers Collections of the Nantucket
Historical Association.

DOCUMENT 6

Josiah Franklin's Bill to Madam Shrimpton,[1] 1706

(By Permission of the Massachusetts Historical Society)

1706		Mad^m: Shrimpton Dr		
Oct	24	To 1 doz Candles	. . . .	00//06//00
No:	28	To 2^E ditto	. . . .	00//01//00
Dec:	22	To 2^E ditto	. . . .	00//01//00
Jan:	8	To 2^E Tallow at 4^d ½	. . . .	00//00//09
	15	To ½ doz Candles	. . . .	00//03//00
Feb:	16	To 1 doz & ½ ditto	. . . .	00//09//00
	27	To a box of candles being 103 10 Torn 93 Neet at 5^d ½	. . . .	02//02//07½

£ 03//03//04½

ℱ Josaih ffranklin

---

[1]Elizabeth Roberts Shrimpton, widow of the late Council member Samuel Shrimpton.

DOCUMENT 7

Josiah Franklin's Receipt from Madam Shrimpton,

February 14, 1707/8

(By Permission of the American Philosophical Society)

1707/8   feb:14

Then Rec^d of M^{dm}. Shrimpton the Sum of three pounds10^d being in full.

ρ Josiah ffranklin

DOCUMENT 8

Josiah Franklin's Bill to Boston for Candles to the Battery, 1710

(By Permission of the Historical Society of Pennsylvania)

1710 Town of Boston Dr to Josiah ffranklin by Candles dd to Battery

May	9	To 2^E	Candles -- -- -- -- -- 00//01//03
	13	To 3^E	ditto -- -- -- -- -- 00//02//00
June	8	To 2^E	ditto -- -- -- -- -- 00//01//04
	18	To 3^E	ditto -- -- -- -- -- 00//02//00
July	4	To 5^E	ditto -- -- -- -- -- 00//03//04
aug	12	To 5^E	ditto -- -- -- -- -- 00//03//04
Sep	9	To 2^E	ditto -- -- -- -- -- 00//01//04
	22	To ½ doz	ditto -- -- -- -- -- 00//04//00
Oct	23	To ½ doz	ditto -- -- -- -- -- 00//04//00
No	22	To ½ doz	ditto -- -- -- -- -- 00//04//00
Dec	26	To 2^E	ditto -- -- -- -- -- 00//01//04
Jan	1	To 4^E	ditto -- -- -- -- -- 00//02//08
	24	To ½ doz	ditto -- -- -- -- -- 00//04//00
feb	24	To ½ doz	ditto -- -- -- -- -- 00//04//00
Mar	21	To 5^E	ditto -- -- -- -- -- 00//03//04
may	1	To 5^E	ditto -- -- -- -- -- 00//03//04
June	26	To 5^E	ditto -- -- -- -- -- 00//03//04
July	9	To 1^E	ditto -- -- -- -- -- 00//00//08
	21	To 2^E	ditto -- -- -- -- -- 00//01//04
aug	5	To 3^E	ditto -- -- -- -- -- 00//02//00
	28	To 5^E	ditto -- -- -- -- -- 00//03//04
Sep	25	To ½ doz	ditto -- -- -- -- -- 00//04//00
No	27	To 3^E	ditto -- -- -- -- -- 00//02//00
Dec	19	To 3^E	ditto -- -- -- -- -- 00//02//00
	30	To ½ doz	ditto -- -- -- -- -- 00//04//00
feb	1	To ½ doz	ditto -- -- -- -- -- 00//04//00

£ 03//11//11

DOCUMENT 9

Josiah Franklin's Bill to Boston for Candles to the South Battery, 1711

(By Permission of Beinecke Library, Yale University)

1711 South Battery Dr To Josiah ffranklin

mar	6	To ½ doz	Candles -- -- -- --	00//04//00
ap:	6	To 5^E	ditto -- -- -- --	00//03//04
	29	To 5^E	ditto -- -- -- --	00//03//04
May	30	To 5^E	ditto -- -- -- --	00//03//04
July	6	To 5^E	ditto -- -- -- --	00//03//04
aug	6	To 5^E	ditto -- -- -- --	00//03//04
Sep	1	To ½ doz	ditto -- -- -- --	00//04//00
Oct	6	To ½ doz	ditto -- -- -- --	00//04//03
Dec:	1	To ½ doz	ditto -- -- -- --	00//04//03
feb	3	To ½ doz	ditto -- -- -- --	00//04//06
mar	6	To 5^E	ditto -- -- -- --	00//03//09
ap	9	To ½ doz	ditto -- -- -- --	00//04//06
may	2	To 1^E	ditto -- -- -- --	00//00//09
	18	To 4^E	ditto -- -- -- --	00//03//00
	30	To 2^E	ditto -- -- -- --	00//01//06
June	9	To 3^E	ditto -- -- -- --	00//02//03
July	2	To 5^E	ditto -- -- -- --	00//03//09
	29	To 5^E	ditto -- -- -- --	00//03//09
aug	27	To ½ doz	ditto -- -- -- --	00//04//06
Sep	27	To 1^E	ditto -- -- -- --	00//00//09
Oct	6	To ½ doz	ditto -- -- -- --	00//04//06

£ 03//10//08

℘ Josiah ffranklin

DOCUMENT 10

Peter Sergeant's Deed of Sale to Josiah Franklin, January 25, 1711/2

(Suffolk Deeds: 26: 108)

Sergeant Esq[r].
et Ux. to
Franklyn

To all People unto whom this present Deed of Sale shall come. Peter Sergeant of Boston in the County of Suffolk within Her Majesties Province of the Massachusetts Bay in New England Esq[r]. and Mehetable his Wife formerly the Wife of Thomas Cooper late of Boston aforesaid Merchant deced and One of the heirs Devisees and Executors of the Last Will and Testament of the Hono[ble]: William Stoughton late of Dorchester in the County of Suffolk afores[d]. Esq[r]. Deced Send Greeting. Know ye That the said Peter Sergeant and Mehetable his Wife for and in consideration of the Sum of Three hundred and Twnety pounds in good Current Bills of Credit within the aforesaid Province to them in hand well and truly paid at and before the Ensealing and delivery of these presents by Josiah Franklyn of Boston aforesaid Tallow Chandler, the receipt whereof to full Content and Satisfaction they the said Peter Sergeant and Mehetable his said Wife Do hereby acknowledge, and thereof and of every part and parcel thereof for themsevlves their heirs Executors and Admin[rs]. Do acquit Exonerate and Discharge the said Josiah Franklyn his heirs Executors Admin[rs]. and Assignes and every of them forever by these presents, As also for divers other good causes and considerations them hereunto moving: they the said Peter Sergeant and Mehetable his Wife have given granted bargained Sold aliened Enfeoffed Conveyed and Confirmed, and by these presents for themselves and their heirs Do fully freely clearly and absolutely give grant bargain Sell aliene Enfeoffe Convey and Confirm unto him the said Josiah Franklyn his heirs and Assignes forever All Those their houses and Tenements with the appur[ces]. And all the Land whereon they stand, and is thereunto belonging and Adjoyning Scituate Lying and being in Boston aforesaid, butted bounded and measuring as Followeth Viz[t]. At the Front or Easward End by Union Street so called, measuring there in breath Thirty eight feet or thereabout, On the Northward side by Hannover Street so called, Measuring there in Length Ninety three feet or thereabout, On the Rear or Westward End by Land formerly of Josiah Cobham deced in the present Tenure and Occupation of Joseph Smith Sadler where it measureth in breath Twenty three feet five Inches or thereabout, and on the Southward side by Land formerly the said Cobham's, and the house and Land formerly apportaining to John Cotta now wholly on this side the Inheritance of the heirs of Thom[s]. Bridge late of

Boston aforesaid Marriner deced where it measureth in Length Eighty seven feet or thereabout. Together with the Shops Ediffices buildings and fences standing thereon, well therein, ways Easments Waters water courses profits priviledges rights Commodities hereditaments Emoluments and appur^{ces}. whatsoever to the said granted and bargained premisses belonging or in any wise appertaining or therewith now or heretofore used sett Lett Occupyed or Enjoyed accepted reputed taken or known as part parcel or member thereof, And the Revercon and revercons remainder and Remainders rents issues and profits thereof And all the Estate right Title Interest Inheritance use possession property claim and Demand whatsoever of the said Peter Sergeant and Mehetable his said Wife and of either of them of in and to the same premisses and every part thereof with all Deeds writings and Evidences relating thereunto. (All which above granted and bargained premisses were heretofore purchased of John Gill then of Milton since deced by the said William Stoughton and are rightly come unto the said Mehetable in her aforesaid Capacity upon the Division and Settlement made of the Estate of the said William Stoughton among all the heirs Devisees and Executors of the said Stoughton, As by the Articles thereof upon Record bearing date on or about the Seventeenth day of July which was in the year of Our Lord 1704. reference whereto being had more fully may appear.[)] To Have and To Hold all the above & before mentioned granted and bargained premisses with the appur^{ces}. & every part and parcel thereof unto the said Josiah Franklyn his heirs and Assignes forever. To his and their own Sole and proper use benefit and behoof from henceforth and forever more absolutely without any manner of Condition redemption or revocation in any wise. And the s^d. Peter Sergeant and Mehetable his said Wife for themselves their heirs Executors and Admin^{rs}. Do hereby Covenant promise and grant to and with the said Josiah Franklyn his heirs and Assigns in manner and form Following. That is to say That at the time of the Ensealing hereof, and until The delivery of these presents they the said Peter Sergeant and Mehetable his said Wife are True Sole and lawful Owners of all the aforebargained premisses, and stand lawfully seized thereof in their or One of their own proper right of a good Sure and Indefeasible Estate of Inheritance in Fee Simple. Having in themselves or One of them full power good right and lawful Authority to grant Sell Convey and Assure the aforesiad premisses unto the said Josiah Franklyn his heirs and Assigns in manner and form aforesaid, and according to the true Intent and meaning of these presents, And that the said Josiah Franklyn his heirs and Assignes shall and may by force and virtue of these presents from time to time and at all times forever hereafter lawfully peaceably and quietly have hold use Occupy possess and Enjoy the abovegranted and bargained premisses with the appur^{ces}. without the lawful or Equitable Lett Suit molestation Eviction Ejection or Interruption whatsoever of them the said Peter Sergeant and Mehetable his said Wife or of either of them their or Either of their heirs or Assignes or of the heirs of the said Thomas Cooper or of any other person or persons whatsoever. And That Free and clear and clearly Acquitted Exonerated and Discharged of and from all and all manner of former and other gifts grants bargains Sales Leases Releases Mortgages Joyntures Dowers Judgments Executions Entails fines Forfeitures Seizures amerciaments Rents Estates rights Titles troubles charges and Incumbrances whatsoever. And Further That they the said Peter Sergeant and Mehetable his said Wife for themselves their heirs Executors and Admin^{rs}. and every of them Do thereby Covenant and grant to Warrant and Defend all the abovegranted and bargained premisses with their appur^{ces}. unto the said Josaih Franklyn his heirs and Assignes forever against the lawful claims and Demands of all and every person and persons whomsoever. And at any time or times hereafter on reasonable request or Demand when need -p -p -p - so required to give and pass unto the said Josiah Franklyn his heirs and Assignes at his and their proper charges; such further and ample Assurance and Confirmation of the premisses as in Law or Equity can or may be reasonably devised advised or required. In Witness whereof the said Peter Sergeant and Mehetable his said Wife have hereunto sett their hands and Seals the Twenty fifth day of January Anno Domini 1711/12. In the Tenth year of the Reign of Our Sovereign Lady Queen Anne over Great Britain &c. Peter Sergeant and a Seal, Mehetable Sergeant and a Seal, Signed Sealed and Delivered in the presence of us Hannah Ellis Eliezer Moody Sen. Received the day and year first withinwritten of the withinnamed Josiah Franklyn the Sum of Three hundred and Twenty pounds in good Curr^r. Bills of Credit in full payment Satifaction and Discharg of the purchase Consideration within Exprest ρ us Peter Sergeant Mehetable Sergeant Suffolk Ss. Boston February 8th. 1711/12. The withinnamed Peter Sergeant and Mehetable his Wife personally appearing before me the Subscriber One of Her Majesties Justices of Peace within the County of aforesaid, acknowledged this Instrument of bargain and Sale to be their Act and Deed, She the said Mehetable freely consenting thereunto. Penn Townsend February 8th. 1711. Received and accordingly Entred and Examined.

ρ Addington Davenport Regist^r:

DOCUMENT 11

Josiah Franklin's Deed of Mortgage to Simeon Stoddard,

February 8, 1711/2

(*Suffolk Deeds*: 26: 109)

Franklyn
et Ux to
Stoddard Esq'.

To all People unto whom these presents shall come. Josiah Franklyn of Boston
in the County of Suffolk within Her Majesties Province of the Massachusetts
Bay in New England Tallow Chandler Sendeth Greeting. Know ye That I the
said Jsoaih Franklyn for and in Consideration of the Sum of Two hundred and
Fifty pounds in good Current Bills of Credit within the aforesaid Province to
me in hand well truly paid at and before the Ensealing and delivery of these
presents by Simeon Stoddard of Boston aforesaid Esq'. the receipt whereof to
full Content and Satisfaction I Do hereby acknowledge and thereof and of every
part and parcel thereof Do Acquit Exonerate & discharge the said Simeon
Stoddard his heirs Executors Admin⁵. And Assinges and every of them forever
by these presents. Have Given granted bargained Sold aliened Enfeoffed
Conveyed and Confirmed, And by these presents for me and my heirs Do fully
freely clearly and absolutely give grant bargain Sell aliene Enfeoffe Convey and
Confirm unto the said Simeon Stoddard his heirs and Assignes forever. All
those my houses and Tenements with the appur^ces. and all the Land whereon
they stand and is thereunto belonging and Adjoyning (which I lately purchased
of Peter Sergeant of Boston aforesaid Esq'. and Mehetable his Wife formerly
the Wife of Thomas Cooper late of Boston aforesaid Merchant deced and One
of the heirs and Devisees of the Last will and Testament of the Hono^ble.
William Stoughton late of Dorchester in the County of Suffolk aforesaid Esq'.
deced) Scituate standing lying and being in Boston aforesaid being butted
bounded and measuring as Followeth That is to say, On the Eastward End by
Union Street so called where it measures in breath Thirty eight feet or
thereabout, On the Northward side by Hannover Street so called, measuring
there in Length Ninety three feet or thereabout, In the Rear or Westward End
by Land formerly of Josiah Cobham deced now in the Occupation of Joseph
Smith Sadler, where it measures in breath Twenty three feet five Inches or
thereabout, and on the Southward side by Land formerly of the said Cobham
and the house and Land formerly of John Cotta, now wholly the possession of
the heirs of Thomas Bridge late of Boston aforesaid Boatman deced where it

measureth in Length Eighty seven feet or thereabout. Together with all and singular the Ediffices Buildings Shops and fences standing thereon well and water therein ways Easments profits priviledges rights Commodities hereditaments Emolum^{ts}. and appur^{ces}. whatsoever thereunto belonging or in any wise apportaining or therewith now or heretofore used Occupyed Enjoyed accepted reputed taken or known as part parcel or member thereof, And the revercon and recercons remainder and remainders thereof, and all the Estate right title Interest Inheritance use possession property claim and Demand whatsoever of me the said Josiah Franklyn and my heirs of in and to the same premisses and every part thereof. To Have and To Hold all the above granted and bargained premisses with their appur^{ces}. And every part and parcel thereof unto the said Simeon Stoddard his heirs and Assignes forever. To his and their own sole and proper use benefit and behoof forevermore. Provided always and it is the true Intent and meaning of these presents and parties to the same any thing herein Contained to the Contrary thereof in any wise Notwithstanding That if I the said Josiah Franklyn my heirs Executors Admin^{rs}. or Assignes shall and do well and truly pay or cause to be paid unto the above named Simeon Stoddard his heirs Executors Admin^{rs}. or Assignes in Boston aforesaid the full and Just Sum of Two hundred Sixty and five pounds in like good Current Bills of Credit within the aforesaid Province at One Entire payment on or before the ninth day of February which will be in the year of Our Lord One Thousand Seven hundred and Twelve/13. without fraud Coven or further Delay, Then this present grant bargain and sale and every clause and Article thereof to be utterly void and of none Effect or Else to abide and remain in full force strength and virtue to all Intents and purposes in the Law whatsoever. And I the said Josiah Franklyn for me my heirs Exec^{rs}. And Adm^{rs}. Do hereby covenant promise and grant to and with the said Simeon Stoddard his heirs and Assignes in manner and form following That is to say, That at and immediately before the time of the Ensealing and Delivery of these presents I the said Josiah Franklyn am true Sole and Lawful Owner of all the abovegranted and bargained premisses with the appu^{ces}., and stand lawfully seized thereof in my own proper right of a good sure and Indefeasible Estate of Inheritance in Fee Simple without any manner of Condition reversion or Limitation of use or uses Whatsoever so as to Alter change Defeat or make void the same. Having in my self full power good right and lawful Authority to grant sell Convey and assure the abovegranted and bargained premisses with the appur^{ces}. unto the said Simeon Stoddard his heirs and Assignes in manner and form aforesaid, and according to the true Intent and meaning of these presents. And that from and after Default made in payment of the aforesaid Sum of Two hundred Sixty five pounds mentioned in the above Proviso. It shall and may be lawful to and for the said Simeon Stoddard his heirs or Assignes quietly and peacefully to Enter into and upon and from thenceforth forever thereafter to have hold use Occupy possess and Enjoy the abovegranted and bargained premisses with the appur^{ces}. Free and Clear and clearly Acquitted Exonerated and Discharged of and from all and all manner of former and other gifts grants bargains Sales Leases Releases Mortgages Joyntures Dowers Judgments Executions Entails Fines Forfeitures Seizures Amerciaments Rents Estates rights Titles troubles charges and Incumbrances whatsoever. And further That I the s^d. Josiah Franklyn for me my heirs Exec^{rs}. and Admin^{rs}. Do hereby Covenant and grant to Warrant and Defend the abovegranted and bargained premisses with the appur^{ces}. unto the said Simeon Stoddard his heirs and Assignes forever against the lawful claims and demands of all and every person and persons whomsoever from and after default made as aforesaid And at any time or times hereafter on reasonable request or demand from and after default made as aforesaid to give and pass such further and ample assurance and Confirmation of the premisses unto the said Simeon Stoddard his heirs and Assignes as in Law or Equity can or may be reasonably Devised Advised or required. In Witness whereof I the said Josiah Franklyn and Abiah my Wife, In Token of her free Consent to these presents and full relinguishment of all her right of Dower and thirds of in and to the abovegranted and bargained premisses have hereunto set Our hands and Seals the Eighth day of February Anno Domini 1711/12 In the Tenth year of the Reign of Our Sovereign Lady Queen Anne over Great Britain &c. Josiah Franklyn and a Seal, Abiah Franklyn and a Seal. Signed Sealed and Delivered in presence of us Peter Sergeant Simon Daniel. Received the Day and year first abovewritten of the abovenamed Simeon Stoddard the Sum of Two hundred and ffifty pounds in good Current Bills of Credit in full discharge of the Consideration above Expressed ρ Josiah Franklyn. Suffolk Ss. Boston February the Eighth 1711/12. The abovenamed Josiah Franklyn and Abiah his Wife personally appearing before me the Subscriber One of Her Majesties Justices of the Peace within the County aforesaid, Acknowledged this Instrument to be their free Act and Deed. = Penn Townsend February 8th 1711. Received and accordingly Entred and Examined

ρ Addington Davenport Regist^r:

Mem°., That on the 28th day of Jan^{ry}. 1722. personally appeared in the Office Simeon Stoddard Esqr. The Mortgagee named in the Deed of Mortgage here Recorded and acknowledged that he had rece^d. of Josiah Franklyn the Mortgagor full Satisfaction for the herein mortgaged p^rmisses, did therfore quitclaim all his right title and Interest therein, and cancelld the Original desiring the Record thereof might be discharged In Witness whereof he at the Same time Subscrib'd his name ___

Teste   John Ballantine Reg^r.   Sim: Stoddard

DOCUMENT 12

Josiah Franklin's Deposition, October 31, 1715

(By Permission of the Boston Public Library)

Octr 31: 1715

Mr Mundane a Stay maker^r Came to Town about 6 weekes agoe an orderly man soe far as I know, $^{he\ Comes}$ *from London* My Brother Benjamin ffranklin a silk dyer from London A sober, diligent labourious man, Came in about 3 week agoe both Lodge at my House

Josiah ffranklin

DOCUMENT 13

Benjamin Franklin the Elder's *Short Account*, June 21, 1717

(By Permission of Beinecke Library, Yale University)

A short account of the Family of Thomas Franklin of Ecton in
Northampton Shire. 21 June 1717.

....

I have a dark Idea of the Granfather of Tho. Franklin my Father, that his name
was Henery, that he was an Atturney and lived at Houghton two Miles from
Northampton, and that he had an Estate there of about Eighty pounds a year
free Land, In his dayes the Gent[s] of that town were for having their Land
Inclosed, but this Gent. being of honest principles, knowing the Laws in force
against it, and that it would be a great wrong to the poor, opposed their
designe & stood in deffence of the poor Inhabitants and soe spent his Estate
and did himselfe nor them any good therby for Might overcame Right. Having
thus spent his Estate his son was put to a blacksmith and setled at Ecton but
whether the Atturney had any other child I know not and therfore I proseed to
speak of

Franklin who lived at Ecton 4 Miles Eastward of Northampton and
practiced the trade of a blacksmith there. The character which I have heard
rep[r]sents him as a reserv'd, unsociable m̄ He lived in his own free hold and
there he dyed, He had onley one son Named Thomas which he bro't up to his
own trade and upon his Fathers demise He was possessed of his free hold at
Ecton aforesaid It being in vallue about 18 or 20*l* ℘ *an*:

Thomas Franklin (the son of Franklin above) was born at Ecton in
Northamptonshire on the 8 day of Oct: in the year 1598. After his fathers
decease He married Jane white Neece to Cott: White of Nethrop near Banbury
in oxfordshire, she had one brother who lived at Grundon two miles from
Ecton and one sister who married m[r] Ride in Warwickshire, and had by him
one son Named Samuel Ride to whom he left about 60*l* per An: free land
which he in a few years spent & sold and became a Gent. Servant and afterward
a Labourer in building the city of London after that Great and Dreadful fire
that burri[d] it on the 2, 3, and 4 of sept 1666, which did destroy 13300 houses,
This samuel Ride had onley one daughter who married a butcher in Clare
Market, Westminster

Thomas Franklin had by his Wife Jane White Nine children whose names
follow

Thomas   born   3 Mar. 1637

Samuel	born	5 Nov. 1641	
John	born	20 Feb. 1643	dyed 7 June 1691
Joseph	born	10 Oct 1646	
Benjamin		20 Mar. 1650	
Hannah	..	29 Oct. 1654	
Josiah	born	25 Dec. 1657	

They had two sons more, Twins, born befor benjam. but tis tho't they dyed unbaptized because their names are not found in the church Register at Ecton where we were all born & brought up.

I remember nothing of my mothers father nor my fathers father but what you have above of my Father it sd he was naturaly of a chearfull temper, pleasant conversation, Just in his dealings, My Mother had a brother White whom I never saw but once. his dealings, as to his trade an Exelent workman, a man of understanding, in the best things, a constant Attender on ordinances, and family dutys, and a great observer of the works of providence. His son Josiah resembles him most both as to his person (save in his feet) and his Naturall disposition. while he was young he made a clock which went well for many years in my remembrance, he alsoe practised for diversion the trade of a Turner, a Gun-Smith, a surgeion, a scrivener, and wrote as prety a hand as ever I saw. He was a historian, and has some skill in Astronomy and chymistry which made him acceptable company to m^r John Palmer the Arch-Deacon of Northampton.

He left off all busines both as to his trade, and farm, sev^r. all years before he dyed. put it into his son Thomas's hand, and went to his son John at banbury in oxfordshire where he continued to the day of his death, which fell out on the 21 of March 1681, being 83 years & 5 months and 9 dayes old. he was of a brown complexion, comly countenance, Inclin'd to corpulency, and had very little hair and used to wear a cap, and a lover of good men of all denominations, He was most Inclin'd to the presbiterian Government and discipline, but when charles 2^d return'd he went to church alsoe for peace and order sake.

His Wife Jane (my Mother whose name as much as his) I shall ever love and Hon^r, was a tall fair comly person Exact in her morals, and as she was Religiously Educated & alsoe Religiously Inclin'd and kept up a Thursday meeting of her godly woman Neighbours, In which they spent in prayer, conference, and repetion of the foregoing Lords day ser*m*, and singing, about 2 hours time which things I being a child and admitted into their company, had the greater opportunity to know. Two things with relation to this I yet remember, her singing the 4 psalm in the old metre for we then know no other, and her speaking with a great deale of pleasure and pressing with a great deal of Earnestness the meditation of the 3 last verses of the 3 chap. of Malachy upon them that were p'sent. And I remeber once she severly chid me for my Backwardness in learning the Lords prayer, and sd If I went to Hell I should there remember that she had warn'd me of my danger and told and Instructed me in my duty.

She had a Long time of Languishing I think I have heard my father say nere seven years und'. that flattering Lingering distemper, a consumtion, which with other Afflictions she bore with much christian patience and resignation, she dyed and was buried in Ecton church-yard on the North-east side, about 4° November 1662 or 3.

We have a Saying in the family, that My Father was twenty years older than my Mother, My Mother 20 yrs older than my brother Thomas, B'. Thomas was 20 years older than my Brother Josiah. and I think it near the matter. She was neat huswifly and Industrious woman and her person most resembled my Neece Sarah of any that I know among all our relations.

Thomas about the year 1665 perswaded my father to let his Land and leave off Husbandry which in a year or two after he did, and his own trade alsoe, and boarded with him for a while but his Temper being passionate did not sute wth. my fathers and so he went to banbury. [afores^d ?] and for some time before bro. Tho. kept a school and sold tobacco, but when his busine ___ of writing Bills, and bonds, and Deeds et ___ was Increased he left off his schoole to Samuel Roberts his neighbour. and so his r'. business still Increasing he at length became a Noted scrivener, and having the advantage of the Arch Deacon and Esq' Catesby two rich mens purses at comand he raised an Estate of about two Thousand pounds. He had by Elenor his wife one Daughter only, her name is mary, she was married to m' Richard Fisher of Welingborough in Northampton shire who has sould all she had at Ecton, that was left by her father and mother. Thomas dyed at Ecton ∧ $^{\text{in comission for receiving the Land tax for the King in}}$ $^{\text{that country.}}$ on 6 Jan'. 1702. and his widdow dyed there about 10 years afterward. He was a black thin man of very mean appearance, but of great understanding and quick app'hension, very passionate, soon reconciled, & Just in his dealings, Highly for the church of Eng. yet wanted a cordial love for its ministers and toward his end had almost turn'd dissenter.

Samuel, was put Aprentis for m' Wilkinson a silk weaver in maid lane southwark, He dyed in the time of his Aprentiship. of a dropsy. He is sd to be very Ingenious, the most comly person in the family and religiously Inclin'd. He and John Loved one the other Entirely Insomuch that when one had done a fault the other would plead, and procure his pardon before he came in sight. He was burried in s'. mary Overy. (alis) s' saviours southwark church yard about the year 1659 or 60.

John served his time to m' Glover a cloth Dyer at 3 cranes in Thames street London. and being Importun'd to set up in the country by m' Warren of Warmington weaver, and not having his health in the city, he setled at Banbury in oxfordshire, he lived a batchelor long and was a sutor to many young women whose love he seldom miss'd of gaining, but then some trifle or other turn'd his affections from them and I tho't he did not fairly leave them. at last he married m^{rs} Ann Jeffs of marson in warickshire with whom he had about 250l by whom he had severall children whose names follow in order

Thomas,   born on 15 Sept 1683.
Hannah
Ann . .   They were all born at Banbury and when their
Mary . .   father and mother dyed, they became the
Jane . .   Care of my bro. Thomas.
Elenor .

While John lived in the city he was as a father to me and helped me thro' my troubles with my m^r Pratt to whom I served 5 years of my time. He was of a very pleassant conversation, could sute himselfe to any company, and did when he pleased, Insinuate himselfe into the good opinion of persons of all qualities and conditions, in marrying he miss'd it as to one maine designe he aymed at, which was the having a Wife that would Assist in his business, but she proved neither capable nor carfull in that point and soe B^r Thomases prediction was in a great measure verified, who once in my hearing, reproving him for courting and for such little causes leaving soe many whose Affections he had gained, Told him that it would return upon him, that he would be met with, and take up with the Worst at last and Indeed soe he did, according to his own confession when he thus Express'd himself. If my Wife was but like other women, If she was but like my sister Benjamin (that is to say my wife) I should ever Adore her, but then he chekt himselfe and s^d: but, may be, It is best it should be as it is, for I should a been apt to set her in the first place. He was a dyer, lived in good repute at banbury many years. the cause of his death was a boyle or sweling which came by a hurt which he got in mounting his horse, It being in his privities (and thinking to keep it secret) he ˄°pened it with a Needle before it was ripe which caused it to Gangrene up into his body it killed him in 3 dayes time, He dyed I think in June 1689. much lamented of rich and poor in Banury for he was a peace maker and a frind to the poor

His son Thomas is a dyer, lives at Lutterworth in Leicester-shire, Han^a. and mary are at London in service there, the other three are Lacemakers and live, Ann at Hartwell with a Gingerbread baker, Jane with Davis blunt at Ashton, and Nelly at m^r Davis's a farmer at wardon, Hannah lived at m^r Keat a banker's near Hungerford market in the strand, Westminster, at the unicorn. Tho. is married and has one son, the 3 towns above named I take to be in Northampton shire.

Joseph was a carpenter, served his time wth m^r Titeomb Just without moregate London. that being one of the city Gates he helpt to build it, His time of Ap^{r}ntis being Expired, m^r cogshall a suffolk Gent. took him down to Aldborough to build him an house which when he had finished he went and setled at Knatshah a town 6 miles distant from alboro. in the same county, where he married sarah Sawyer, Daughter of m^r Saw⌃ers toward to . . . . . . He had by her one son named Joseph, born after his fathers death. He was facetious in his comon conversation, his Judgm^t was for the Church of England, but his wife was otherwise Enclin'd. He dyed on st Andrews day, the 30 Nov. 1683. she married againe to m^r Blackmare at blybrow near Dunwick in Suffold, and there her son Joseph dyed about 21 year of his Age.

Benjamin served his time to m^r Pratt 5 years, m^r Paine 2 years. Dyers of skeyn silk black. which he practised for about 7 years after. and then learned to dye skeyn silk into collours, that he followed for about Eleven years more, then turn'd Ragg dyer as tis called in London, that is dying wro't silk in the peece and when made into Garments, this he did for about seventeen years, but not having the desired success he left off and went to New England and Landed at Boston on 10 Oct. 1715

Before this Benjamin there were born two sons more, Twins, which as I s^d (tis tho't) dyed unbaptiz'd. In the year 1683 on fryday 23 Nov. He married Mrs Hannah Welles, Daughter of Mr Samuel Welles minister of Banbury in oxfordshire, this m^r W. was one of those 2000 that were turn'd out soon after King Charles 2^d restoration, on 24 Aug 1662 comonly called, Black Bartholomew day.[1]   Dr Mather is another of them, who in England are called Dessenters, together with those that follow them, Benj. had By his wife Hannah Ten children, Namely

Samuel	born on	15	Oct.	1684	
Benjamin }		6	Aug.	1686	dyed 22 Apr 87
Jane . { dead . . .		14	feb.	1687	
Hannah . |		18	Nov.	1689	dyed 31 Dec. 1710
Thomas . J		31	Aug.	1692	dyed 2 mar: 94

---

[1]He also wrote,

> M^r. Sam: Welles M.A. was minister of Banbury in Oxford shire until the Black Bartholomew, when he was Ejected with about 2000 Ministers more, He lived afterw^d and dyed in his own house there, about the year 1678: Dorothy his Widow removed to Londo in 1682. B.F. married her Daughter Hannah the youngest of 12. on the 23 Nov. 1683. And my son sam: was born in White-chappel parish in Goodms fields.15 Oct. 1684.

Commonplace Book, Publications of the Colonial Society of Massachusetts (Boston, 1907), vol. 10, p. 205.

Elizabeth	27	Oct.	1694	
Mary . . }	23	Apr.	1696	dyed 27 Aug 96
John . . { dead . . .	8	Apr.	1699	
Joseph . |	27	Jan^y.	1700	
Josiah . J.	3	Jan^y.	1703	dyed 10 Jan^y:

Hannah the Mother of these dyed in princes street in st Ann's parish in Westminster on the 4° Nov. 1705. and in her I Lost the delight of mine Eyes, the desire of my heart, and the comfort of my life. she wrote severall things for her own private use, some of them are in her son Samuels hands. Hannah My Daughter was of a weakly constitution, as was her mother, and took after her as to writing, rediness of witt and curious working with her Needle, but was not soe happy in her Natural temper which was somwhat like her fathers. which he was apt to Impute to her sickly dispostion when I had begun too much to set my Affections on her, as standing in her mother stead, and in good measure filling up her room, having a good understanding in the best things ⌃ O dismal change for Father to reherse, his Daughter Turn'd unto the Just reverse. of a discreet deportment toward others, and pruden^t. houswifely neat and saving in all her managements. It pleased the holy God to take her (I hope) to himselfe on 31 Dec. 1710. Bet: 11 and 12 at Night being the last day and the last hour of the old year her brother sam. has somthing that she wrote alsoe for her own private use

Samuel who was born on wed. at 8 a clock in the Evening in prescote street in Goodm^s fields was baptiz'd by m^r James who used to preach near Nightingall lane near well closs. who on that occation did Exelently open and Apply that text in the 20 Ezek. 37 I will cause you to pass under the Rod et.

Elizabeth was born Near the falcon staires in the parish of christs-church southwark on Saturday. and baptiz'd by m^r Nathanael Vincent It was s^d of her by m^rs pelly a Docteress that betty was short lived she is of a good temper and meek spirit yet knows how to resent an Injury, she is of a healthfull constitution and is near the age of my Neece m^rs mary Homes

It pleased God to take away all the rest of my children in their Infancy, none Except Ben. & Thomas lived twelve months if they lived soe long which I am not certain of.

Hannah, of whom my father used to say, when any asked how many children he had, I have had seven sons, and they have every one a sister she had several good offers, but as She was a hinderance in brother Johns closing with several good offers, soe she herselfe refused severall and took up with what proved the worst she was married to John Morris son of Billing Morris of Ecton aforesaid who had with her 100*l* He was a zealous son of the church and made her soe, he had good and profitable busines of his trade which was a rag dyer but his fancy lead him to building wherby he Involved himselfe, and soon after dying he left his widdow 3 Daughters and six hundred pounds in Debt out of which she never got all her life long. but dyed in debt to all and more than She dealt with {24 June 1712.

He dyed 17 June 1695. and left 3 Daughters whose names are as follows but their age I doe not know, they are all single and live in London. she Dyed 24 June 1712.

Elenor   . Has a charming tongue, is of a very obliging cariage free in her promises but far from endeavours to perform them.

Jane   . is of few words, and many deeds, yet guilty of the above named fault these two speak and write and read french near as fluently as English, are redy witts and highly for the church of England

Hannah   . is of very few words you must draw them out, or goe without them of a bushfull countenance and a weak constitution, they are all 3 of very smal Appetites, I know some one woṁ that would eat more than they all, they did all together follow the dying silk garments and scouring since their mothers death which happened 24 June 1712. but now they all goe to Service.

Josiah was a dyer, served his Aprentiship to his brõ. John at banbury where he married Ann Child of Ecton the Daughter of Robert Childe there. but things not succeeding there according to his mind, wᵗʰ. the leave of his frinds and father he went to New England in the year 1683. in order to which voyage he was come up to London at the time when the Noble Lord Russel was murder'd

He had by his wife Ann seven children.

Elizabeth born ..	2 Mar. 1677 ⎱
Samuel ... dead ..	16 May 1681 ⎰ these born banbury
Hannah	25 May 1683 ⌡
Josiah ... dead ..	23 Aug 1685
Ann	5 Janʸ 1686
Joseph ⎱	6 feb. 1687
⎰ Dead	
Joseph 2ᵈ	30 June 1689

By his 2ᵈ wife Mʳˢ Abiah Foulger

John born ..	7 Dec. 1690
Peter	22 Nov. 1692
Mary	26 Sept 1694
James	4 feb. 1696
Sarah	9 July 1699

Ebenezer   . . . . .   20 Sept 1701

Thomas } . dead   7   Dec. 1703

Benjamin   6   Jany. 1706

Lidia   8   Aug 1708

Jane . . . .   27 Mar 1712

    After he came to Boston in N. England, he made several Essays, in several sorts of bussines, and at last fixed upon the trade of Tallow chandler, and Sope maker, in which it has pleased God so to bless his diligence and Endeavours, that he has comfortably bro't up a Numerous family, providing for and disposing off almost halfe of them in a credible maner and himself lives in a good repute among his frinds and Neighbours at the blue ball in union street Boston the place where this briefe account was writen on the 1. 2. 3. Of July 1717 by his brother

                  Benj. Franklin

DOCUMENT 14

Ambrose Vincent's Bill to David Stoddard, 1718-19

(By Permission of the Massachusetts Historical Society)

1718	M^r. David Stoder^d.	D^r
		l
Sept 27.	To Scowering a flowerd Damask Gound	) 0 : 6 : 0) ^ES
	To Scow: a white and Gold Colour Stript Latten gound	) 0 : 6 : 0) ^ES
Oct o: 15.	To Dying 2 gounds organe Colour.	0 : 12 : 0 DS
1719		
March 30.	To Scow a rich Damask gound . . . . . and a breath of a peticoat)	) 0 : 7 : 0 ES
	To Scow: a yallow lineing	^ES0 : 6 : 0
April 25.	To Sundrys. as p^sd note of Mad^m.. Beach) Sent inn . . . .)	1: 16 : 0

                                £3:13: – £3 :15 : 0

                     Ambrose Vincent

Document 15

Jeremiah Bumstead's Bill to David Stoddard, February 21, 1722

(By Permission of the Massachusetts Historical Society)

Boston feb: 21._____1722

Mʳ David Stoddard to Jeremiah Bumstead Dʳ

1720___march 24 to a hand=Saw Sett & Sharp - - - - - - - - - -	0=:1=0
Decem: 2 to a Crosscutt=Saw 2ˢ & hand=Saw Sharpt 1ˢ	0=:3=0
1721___October 16 to a new hand=Saw 2/6 Decem: 8 the Crosscutt 2ˢ	0=:4=6
1722. feb: 6 to a Crosscutt=Saw Sett & Sharp - - - - - - - -	0=:2=0
	———
	0=10=6
	———

DOCUMENT 16

Josiah Franklin's Deed of Mortgage to Hannah Clarke, January 29, 1722/3

(*Suffolk Deeds*, 36: 191)

Franklin Ux
to
Clarke

To all people unto whom these presents shall come. Josiah Franklin of Boston in the County of Suffolk within His Maj[ties] Province of the Massachusetts Bay in New England Tallow Chandler, sendeth Greeting Know yee, that I the s[d] Josiah Franklin for in Consideracon of the Sum of Two hundred & twenty pounds in good Curr[r]. bills of Credit on the Province afores[d]. To me in hand well & truly paid at and before the Ensealing & delivery of these p[r]sents by Hannah Clarke of Boston afores[d] Widow the receipt whereof to full Content and Satisfaction I do hereby acknowledge and thereof and of every part and parcell thereof do acquit Exonerate & discharge the s[d]. Hannah Clarke her Heirs Exec[rs]. Adm[rs]. & Assignes & every of them forever by these presents Have given granted bargained sold aliened enfeoffed conveyed and confirmed and by these p[r]sents for me & my Heirs Do fully freely clearly & absolutely give grant bargain sell aliene enfeoffe convey & confirm unto said Hannah Clarke her Heirs and Assignes forever All Those my Houses & Tenements with the appur[ces]. & all the Land whereon they stand & is thereunto belonging and adjoyning (which I purchased of Peter Sergeant of Boston afores[d] Esq[r]. dece[d]. and Mehetabel his Wife, formerly the Wife of Thomas Cooper of Boston afores[d]. Merch[t]. Dece[d]. and One of the Heirs and Devisees of the Last Will & Testam[t]. of the Hon[ble]. William Stoughton of Dorchester in the County of Suffolk afores[d] Esq[r]. dece[d].) scituate standing lying & being butted bounded measuring as Followeth, That is to say, on the Eastward end by Union Street so called, where it measures in breath Thirty eight feet or thereabouts, on the Northward side by Hanover Street so called, measuring there in the length ninety three feet or thereabout in the Rear, or Westward end by Land formerly of Josiah Cobham dece[d]. where it measures in breath twenty three feet five Inches or thereabouts and on the Southward side by the Land formerly of the s[d]. Cobham & the house and Land formerly of John Cotta, where it measureth in length eighty seven feet or thereabout, Together with all and singular the edifices buildings shops and fences standing thereon Well and Water therein ways Easem[ts]. profits priviledges rights Commodities hereditam[ts]. emolum[ts]. & appur[ces]. whatsoever thereunto belonging or in any wise apportaining or therewith now or heretofore used occupied enjoyed accepted reputed taken or known as part parcell or memebr thereof And the Revercon & Revercons remaind^r. and Remainders thereof And all the Estate right title interest Inherit^a. use possession property claim & demand whatsoever of me the s^d. Josiah Franklin & my Heirs of in & to the same p^{r}mises and every part thereof To have and to hold all the abovegranted & bargained p^{r}mises with their appur^{ces}. And every part & parcell thereof unto the s^d. Hannah Clarke her Heirs & Assignes forever to her & their own sole & proper use benefit & behoofe forevermore Provided always, and it is the true intent and meaning of these presents and parties to the same any thing herein Contained to the Contrary thereof in any wise notwithstanding, that if I the s^d. Josiah Franklin my Heirs Exec^{rs}. Adm^{rs}. or Assignes shall & do well and truly pay or cause to be paid unto the abovenamed Hannah Clarke her Heirs Exec^{rs}. Adm^{rs}. or Assignes in Boston afores^d. the full & just Sum of Two hundred & twenty pounds in good & Lawfull publick bills of Credit on the Province afores^d. with Lawfull Interest for the same on or before the twenty eighth day of Jan^{ry} next ensuing the day of the date hereof which will be in the year of Our Lord One Thousand Seven hundred & twenty three without fraud Coven or further delay, then this present grant bargain & sale & every Clause and article thereof to be utterly void & of none effect or else to abide & remain in full force strength & virtue to all interests & purposes in the Law whatsoever And the s^d Josiah Franklin for me my Heirs Exec^{rs}. and Adm^{rs}. do hereby Covenant bargain & agree to & with the s^d. Hannah Clark her Heirs & Assignes in manner and form following, That is to say, that at & immediately before the Ensealing and delivery of these presents, I the s^d. Josiah Franklin am true sole and Lawfull Owner of all the abovegranted & bargained p^{r}mises with the appur^{ces}. & stand lawfully seized thereof in my own proper right of a good sure & Indefeasible Estate of Inherit^{ds}. in Fee simple without any manner of Condition & Revercon or limitacon of use or uses whatsoever so as to alter change defeat or make void the same. Having in my self full power good right & lawfull Authority, to grant sell convey & assure the abovegranted & bargained p^{r}mises with appur^{ces}. unto the s^d. Hannah Clarke her Heirs & Assignes in manner & form afores^d. and according to the true intent & meaning of these p^{r}sents, and that from and after default made in paym^t. of the afores^d. Sum of Two hundred & twenty pounds with Interest on the same mentioned in the above proviso, it shall & may be lawfull to and for the s^d. Hannah Clarke or Heirs or Assignes quietly & peaceably to enter into & upon and from thenceforth forever thereafter to have hold use occupy possess & enjoy the above granted & bargained p^{r}mises with the appur^{ces}. free and clear & clearly acquitted exonerated & discharged of and from all & all manner of former other gifts grants bargains sales leases releases mortgages joyntures dowers Judgm^{ts}. Executions entails fines forfeitures seizures amerciam^{ts}. rents estates rights titles troubles Charges and Incumbrances whatsoever, And further that I the s^d Josiah Franklin for me my Heirs Exec^{rs}. & Adm^{rs}. do hereby Covenant & grant to Warrant & Defend the above granted & bargained premises with the appur^ces. unto the s^d. Hannah Clarke her Heirs and Assignes forever against the Lawfull claims & demands of all & every person & persons whomsoever from & after default made as afores^d. And at any time or times hereafter on reasonable request or demand (from and default being made as afores^d.) to give pass such farther & ample assurance & confirmacon of the p^rmises unto the s^d. Hannah Clarke her Heirs and Assignes as in Law or Equity can or may be reasonably devised advised or required In Witness whereof I the s^d. Josiah Franklin & Abaih my Wife in token of her free consent to these presents and full relinguishm^t. of all her right of Dower & thirds of in & to the abovegranted & bargained p^rmises have hereunto sett our hands & Seals the twenty eighth day of Jan^ry Anno Dom. One Thousand Seven hundred & twenty two & in the ninth Year of the Reign of Our Sovereign Lord George of Great Britain &c. King Defensers Fidei. Josiah Franklin & a Seal. Abiah ffranklin & a Seal. Signed Sealed & Deliv^d. in the p^rsence of us, the word (meaning) in the Marg^n. of the 2^d. Side, and the word (after) between the lines in the 3^d. side being first added, Erasmus Stevens, Jo Brandon. Received on the day of the Date abovewritten of M^rs. Hannah Clark the Sum of Two hundred and Twenty pounds, being the full Consideracon above expressed ꝑ Josiah Franklin Suffolk Ss. Boston Jan^ry. 29th. 1722/3. The above named Josiah Franklin and Abiah his Wife personally appearing before me the Subscriber One of His Maj^ties. Justices of the peace within the County afores^d. acknowledged this Instrument to be their free Act and Deed, Sam^ll. Chekley. Jan^ry. 30th. 1722. Received and accordingly Entred and Examined.

ꝑ John Ballantine Reg^r.

Memorandum, That on the 9th. Of August 1739, Personally Appeared in the Office, the Honourable Josiah Willard Esquire and Dame Hannah his Wife, late Hannah Clark the mortgagee named in this Deed of Mortgage here Recorded, And Acknowledged they had Receiv'd full satisfaction for the therein Mortgaged Premisses, and did Quit Claim all Right Title and Interest therein, and cancelled the Original Deed, Desiring the Record thereof might also be discharged. In witness whereof they at the same time hereto Subscribed their Names.

Teste   Sammuel Gerrith Reg^r   J. Willard
                                              Hannah Willard

Document 17

James Davenport's & Josiah Franklin's Bond to Josiah Hobbs, May 1, 1723

(By Permission of the Massachusetts Archives)

Know all Men by these presents, that *wee James Davenport Backer & Josiah Franklin Talow Chandler both of boston in the county: Suffolk in new England are* Holden and stand firmly Bound and Obliged unto *Josiah Hobbs of Boston husbandman in the aforesaid Provins* in the full and just Sum of *one hundred pounds in good &* Lawful Money of New-England, to be paid unto the said *Josiah Hobbs: or his heirs: or* Executors, Administrators or Assigns: To the which Payment well and truly to be made, *wee* bind ouer-*selves: our* Heirs, Executors and Administrators firmly by these Presents Sealed with *our*: Seal. Dated the *first:* Day of *may* Anno Domini 1723:

The Condition of this present Obligation is such, That if the above Bounden *James Davenport or Josiah Franklin or their* Heirs, Executors, or Administrators, or any of them, shall and do well and truly Pay, or cause to be Paid unto the said *Josiah Hobbs or his heirs or* Executors, Administrators or Assigns, the full and just Sum of *fifty pounds* in good Bills of Credit on the Province of the Massachusetss-Bay, or Current Lawful Silver Money of New-England, with Lawful Interest for the Same, on or before the *first:* Day of *may next* which will be in the Year of our Lord, One Thousand Seven Hundred and *Twenty four*: without Fraud, Coven, or further Delay; then the above written Obligation to be Void and of none Effect; or else to Abide and remain in full Force and Virtue.

Signed, Sealed and Delivered
in the Presence of us,
James Long   James Davenport [seal]
William Payn   Josiah Franklin [seal]

DOCUMENT 18

Josiah Franklin's Bill to President John Leverett, April 1724
(By Permission of the Massachusetts Historical Society)

Mr Leverett President Dr To Josia Franklin, 1724

1724

ap ..   To 72r Candles at 11 - - - -   03 - 06 - 0
          Recd 37l Suett   at 8d - - - -   01 - 04 - 8
                                                         _____
                                                         2 = 1 = 4

DOCUMENT 19

James Davenport's & Josiah Franklin's Bond

to Josiah Hobbs, April 1,1725

(By Permission of the Massachusetts Archives)

Know all Men by these Presents, That *we James Davenport of Boston in the County: of Suffolk Backer and Josiah Franklin Talow chandler in Boston in said County are joyntly and severally* Holden and stand firmly Bound and Obliged unto *Josiah Hobbs of Boston in aforesaid County husband man* in the full and just Sum of *one hundred pounds* Lawful Money of New-England, to be paid unto the said *Josiah Hobbs or his heirs* Executors, Administrators or Assigns: To the which Payment well and truly to be made, *we* bind *our-selves our* Heirs, Executors and Administrators *or asigns* firmly by these Presents Sealed with *our* Seal. Dated the *first* Day of *aprill:* Anno Domini, 1725 *and in the Eleventh year of the Reign of our Soveraign Lord, Gorge King of Great Britain*

The Condition of this present Obligation is such, That if the above Bounden *James Davenport or Josiah Franklin they or either of their* Heirs, Executors, Administrators, or any of them, Shall and do well and truly Pay, or cause to be Paid unto the said *Josiah Hobbs his heirs* Executors, Administrators or Assigns, the full and just Sum of *fifty pounds* in good Bills of Credit on the Province of the Massachusetts-Bay, or Current Lawful Silver Money of New-England, with Lawful Interest for the Same, on or before the *first* Day of *aprill next* which will be in the Year of our Lord, One Thousand Seven Hundred and *twenty six* without Fraud, Coven, or further Delay; then the above written Obligation to be Void and of none Effect; or else to Abide and remain in full Force and Virtue.

Signed, Sealed, and Delivered
in the Presence of us,
Charls [Harriss?]　　　　　James Davenport
William Payn　　　　　　　Josiah Franklin

Middlesex County Court Warrant, February 13, 1729

Middlesex, Ss. George the Second, by the Grace of GOD, of *Great Britain, France* and *Ireland*, KING, Defender of the Faith, &c.

To the Sheriff of Our County of Suffolk-- his Under-Sheriff, or Deputy, Greeting.
WE Command you to Attach the Goods or Estate of *James Davenport of Boston in the County of Suffolk Baker and Josiah Franklin Tallow chandler in Boston in the said County*-- to the Value of *One hundred Pounds*, and for want thereof to take the Bodys of the said *James Davenport and Josiah Franklin* (if *they* may be found in your Precinct) and *them* safely keep, so that you have *them* before Our Justices of Our Inferior Court of Common Pleas, next to be Holden at *Charlestown* within and for Our said County of *Middlesex*, on the *Second Tuesday* of *March* next: Then and there in Our Said Court to Answer unto--

Josiah Hobbs of Weston in our County of Middlesex yeoman (alias) Josiah Hobbs of Boston in the aforesaid County Husbandman In a Plea of Debt for that the Def on the first day of april 1725 by their Bond of that date in Court to be produced bound themselves in One hundred Pounds lawfull money of New England to be paid to the Pl*: on demand yet tho: often requested have not paid the same but still unjustly detain it--*
To the Damage of the siad *Josiah Hobbs as he saith* the Sum of *One hundred Pounds*; which shall then and there be made to Appear, with other due Damages: And have you there this Writ, with your Doings therein. Witness *Jonathan Remington Esq*; at *Cambridge* this *13th* Day of *February* In the *Third* Year of Our Reign. *Annoque Domini, 1729*

Sam.l Phipps Cler--

DOCUMENT 20

Josiah Franklin's Receipt from James Bowdoin[1], January 26, 1731/2

(By Permission of the New York Public Library)

Boston Janu[ry]: 26[d]: 1731/2

Rec[d]: of M[r]: James Bowdoin Twenty Pounds Cash and Also Received a Bond of him of mine & my Son James Franklin for one hundred & Ninety Six pounds Eleven Shillings & Eight penece also a bond of mine with my said Son James & W[m]: Lowder for one hundred & fifty pounds both Dated, the Seventh Day of June 1720 being in full of all accompts from the Beginning of the world to this Day as witness my hand ----

ρ Josiah Franklin

---

[1]One of the richest merchants in Boston, James Bowdoin I (1676-1747) was from a French Huguenot family. He married Sarah Campbell (1687-1713), the daughter of John Campbell, bookseller and publisher of the *Boston News-Letter*. The marriage was performed by the Reverend Ebenezer Pemberton of the South Church on July 18, 1706. Later a member of the Brattle Street Church, he was the father, with his second wife Hannah Portage, of James Bowdoin II (1726-1790), governor of Massachusetts. Gordon E. Kershaw, *James Bowdoin II: Patriot and Man of the Enlightenment* (Lanham, Md., 1991), pp. 1, 10, 18.

Document 21

Probate Letter and Josiah Franklin's Will, October 20, 1744

(*Suffolk Probate Records*, 44: 221-23)

Probate of
Josiah Franklin's

Edward Hutchinson Esqr. Commissioned by his Excellency William Shurley Esqr: Captain General and Governour in Chief in and over his Majesty's Province of the Massachusetss Bay in New England by & with the Advice and Consent of the Council to be Judge of the Probate of Wills, and for Granting Letters of Administration on the Estates of Persons deceased, having Goods, Chattles, Rights or Credits in the County of Suffolk within the Province aforesaid. To all unto whom these Presents shall come, Greeting. Know Ye, That upon the Day of the Date hereof, before me, at Boston, in the County aforesaid, the will of Josiah Franklin late of Boston aforesaid Tallow Chandler deced, to these Presents annexed, was proved, Approved and Allowed, who having while he Lived, at the Time of his Death, Goods, Chattels, Rights or Credits in the County aforesaid; and the Probate of the said Will and Power of committing Administration of all and singular the Goods Chattels, Rights and Credits of the said deced, by Virtue thereof appertainning unto Me, The Admcon of all and singular the Goods, Chattels, Rights and Credits of the said deced, And his Will in any manner concerning, is hereby committed to his Wife Abiah Franklyn and his Son John Franklin Executors in the same Will named, Well and Faithfully to Execute the said Will, and to Administer the Estate of the said deced according thereunto: And to make a true & perfect Inventory of all and singular the Goods, Chattels Rights and Credits of the said deced; And to Exhibit the same into the Registry of the Court of Probate, for the County aforesaid, at or before the Seventh Day of November next ensuing: And also to render a plain and true Account of their said Adminacon upon Oath. In Testimony whereof, I have hereunto set my Hand & the Seal of the said Court of Probate, Dated at Boston the Seventh Day of August, Annoque Domine, 1750.

Edw:^d Hutchinson

Franklin
Josiahs
Will

In the Name of God Amen this twentieth Day of October One thousand seven hundred & forty four I Josiah Franklin of Boston in the County of Suffolk Tallow Chandler being of sound disposing Mind and Memory, Do make this my last Will and Testament in Manner following, Viz: First & Principally I Commend my Spirit into the Hands of Almighty God my Faithfull Creator, relying upon his free Grace & Mercy for Eternal Life thro' the redemption which is in his only begotton Son Jesus Christ, And my Body I commit to the Earth to be decently Interred at the Discretion of my Executors hereafter Named, in Faith of the Resurrection thereof, and the Joyfull Reunion of Both Soul and Body at the last Day, and as touching such Temporal Estate as God hath betrusted me withall, I Will and Dispose thereof as follows, That is to say, Imp^{rs}: I Will that my just Debts and funeral Charges be well and truly paid by my Executors And if my Son John Franklin shall Advance any of his own Money for or towards the Discharge of any of the Bonds which I stand bound in, then my Will is he shall be Allowed out of my Estate Lawful Interest for what he shall so Advance or pay towards, or for Discharge of said Bonds from the time of payment thereof untill my Wife's Decease. Item I Give to my loving Wife Abiah Franklin all the Income or Rents of my whole Estate and Goods, And the use of the two Rooms we now live in. Allowing the Lodgers to be in as now it is Used she allowing out of it the Interest that will be due to my Creditors while she lives. Item I Give to my Son Benjamin in New Tenor to the Value of thirty Pounds Old Tenor besides what his Share will be hereafter Exprest, I Will that at my Wife's Decease, there be a just and Legal Aprizement made of House & Land and Goods in Order to a just Division of above s^{d}.. Estate which I Will and Bequeath to my Children & Grand Children in manner & proportion following viz. To my Daughter Elizabeth Dows one single Share or ninth Part of the apprized Value thereof, To my Son Samuel Franklin deced his Children. his Children one single Share or ninth part thereof, To my Daughter Ann Harris deceased her Children one single Share or one ninth part of the apprized Value thereof, To my Son John Franklin one single Share or ninth Part, To my Son Peter Franklin one single Share or ninth Part, To my Daughter Holmes deced, and her Children half a Share or ninth part of apprized Value thereof, To my Daughter Sarah Davenport deceased's Children One single Share thereof, To my Son Benjamin one single Share thereof, To my Daughter Lydia Scott one half or ninth Part or single Share, And to my Daughter Jane Mecom One single Share thereof, To my Son James his Children ten shillings apiece New Tenor, & that all the Bonds and Obligations to me of their Fathers to be Cancelled, and what I gave them to be taken out before the abovesaid Division, as Also Son Benjamin Franklin's thirty Pound Old Tenor, The Reason Son Holme's Share is but half a Ninth is because I paid for him I suppose more than a half Share may come to, My Will also is that if my House be destroyed by Fire or any other Accident during my Wife's Life, then she shall be no longer obliged to pay any interest, but it shall be Charged on my Estate, Lastly, I Do hereby Constitute and appoint my Wife Abiah, and my Son John Franklin executors of this my Will, Revoking all former Wills by me Made, In Testimony whereof, I have hereunto set my Hand and Seal the Day and Year First above written. The aforewritten Will was Signed Sealed and Delivered, published & Declared by Josiah Frnaklin to be his Last Will and Testament in the presence of us Jeremiah Bumstead jun[r]: Benjamin Eaton Rich:[d] Barrington.

Josiah Franklin & a Seal

DOCUMENT 22
Tallow Chandler Thomas Clark's Inventory,
May 1, 1749

(*Suffolk Probate Records*, 42: 375-76)

Boston May 1, 1749
The Inventory of the Goods of Thomas Clark of Boston deced are as
followeth viz:

To 1 Great Coat @	£ 8 .. -
To a Short Coat & Jackett of Cloth with Scarlet Lining	25 .. -
To a Camblet Coat £8-../To 2 cloth Coats & Jacketts £5	13 .. -
To a old pair of Breeches & homespun Jackett	2 .. -
To 1 old fastian Jackett 15/ To 6 Old Shirts £8	8 15 -
To 3 Sheets 2 towells & 2 Caps	3 10 -
To 4 pair of Stockings £2- To 1 Desk £5	7 .. -
To 1 pair of Shoe & 3 knee Brekles	
& 1 pair of Buttons all silver	5 .. -
To 2 Second Bottom Bedsteads	5 .. -
To 1 Corded D°. 15/To 1 Chest 40/ To 1 Bible 15/	3 10 -
To 1 Table	2 10 -
To 2 pair of handirons & 1 Gridiron & a tramel & a tostᵉ . . .	2 .. -
To 10 Chairs £4-../ To 4 Old Tables a Cradle & Bellowses	
& base of Bottˢ 20/	8 .. -
To a Tea Kittle & 4 Candlesticks & 2 potts	3 .. -
To 1 Hat & Wig £1-../To one Ax 10/	1 10 -
To 1 Old Feather Bed & the Cloaths of it £15. ../	
To 1 D°: £22	37 .. -
To 11 Boxes of Candles called	110 .. -
To the Boxe for D° called cash	3 17 -
To Kase Tallow 72ˡᵇ @ 3/	10 16 -
To Dryed Tallow 550 wᵗ: @ 3/6	96 5 -
To Candle Rods 120 @ 6ᵈ ℗ Rod	3 10 -
To Table & Frame 2..10/To a small metal Kittle 2..10/	5 .. -
To a candle Mould & Press 6..10/	
To 1 pair of Stillyards 1..10/	8 .. -
To 142 wᵗ: of Babery Tallow @ 7/6 ℗ pound	51.19 -
To 3 pair of Scales & 1 pair of small Stillyards	6 .. -
To ½ Crack of Earth Ware	3 .. -
To the ½ of an old Metal Kittle & ½ of a whole one called . .	5 .. -
To 35 Wᵗ: of hard Soap @ 4/℗ pound	7 .. -

To 1 Soft Dº: & old hogsh^{ds}. & B^{bll}: . . . .   5 .. -

To Old Copper Kittle & Close Stool . . . .   18 .. -

To the one half of an old House & peice of Land . . . . . .   250 .. -

To Candles in the Shop & elswhere . . . .   4 .. -

old Tenor £722-2

This within is the Inventory of the Real & Personal Estate of Thos: Clark of Boston deced taken ꝑ

> James Ramsey
> John M Cleary
> Robt: Campbell

Suffolk Ss: By the Honble: Edward Hutchinson Esqr: Judge of Pro: &c
Jane Clark Executrix presented the within written & made Oath that it Contains a true & perfect Inventory of the Estate of her late Husband Thomas Clark deced so far as hath come to her Knowledge & that if more hereafter appears, she will cause the same to be added. The Subscribing Appraisers were also sworn as the Law directs. ----

Boston May 16th 1749

Edwd Hutchinson

DOCUMENT 23

Inventory of Josiah Franklin's Estate, October 24, 1752

(*Suffolk Probate Records*, 47: 437-38)

Franklin Josiah's Inven^y.

An Inventory of the Estate of M^r: Josiah Franklin late of Boston, taken by us
the Subscribers Octo^r. 24th. 1752.___

The House and Land in Union Street	£253..6..8
10 Silver Porrenger & Spoon 59/2 5 tables 14/8,	
1 Couch 5/4	3..19..2
1 Clock 30/, 2 Desk 16/, 2 looking Glasses 6/8	2..12..8
1 Iron Stove 32/, 2 Easy Chairs 7/7,	
1 high Candlestick 8/	2.. 7..7
12 Chairs 3/2¼, 13 d°. 6/, 5 feather Beds 19../..5	19..10..7¼
5 P Curtains 29/4, 6 Bed quilts 44/.	
1 Bed & 1 Suit of Curt^s. 13/4	4..6..8
11 P^r. Sheets 4..10..8, 5 Bedsteads & Cords 25/4	5..16.___
7 bolster Cases 4/8, 16 pillow Cases 6/4¾,	
7 lable Cloths 11/2¼	1..2..3
7 P^r. of blankets 48/, 4 Rugs 14/, 4 P^r. Bed rods 10/8 . .	3..12..8
1 flock Bed, 8 Straw d°. 1 Chest	2..1..4
1 Chest of drawers, 1 Gun & 2 Swords	13..4
Some old Cloths 21/4, 1 Lanthern 5/4	1..6..8
1 Jack & w^{ts}: 6/8, 1 P^r. handirons, tangs & Shovell 5/4 . .	12.___
1 Warming Pan, 16 Chafindish, 1 P^r. flattirons	6.___
10 Candlesticks, 3 Iron pots, 1 P handirons	.. 13..10
2 Trammels & Sund^y. small things	10..8
2 brass Kittles, 3 brass Skillits	2..4..8
1 Copper 22/, 1 Coffee pot 4/, 1 Iron kittle 3/4	1..9..4
1 flesh fork, 1 Scummer 1/4, tin ware 1/4	2..8
1 Bedspan, 1 tub, 1 tray & Sundry small things	.. 9..10½
3 Powdering tubs, 1 Jar 16/, 1 die tub, 1 wood horse 4/ . .	1.___
1 pe[pp]er mill, 13 Earthen plates, 1 glass Bottle	3..¾
37^a of Dishes & Plates 32/6, 2 large Bibles 34/8,	
1 Concordance 2/8	3..9..10
Mr Willard's Body of Divinity	6..8
A Parcell of small Books	14..8
	£312..18..10½

Mem°: a P^r. Scale & w^{ts}: not apprized, sold
For £1..4..8. A Suit of Cloaths sold
for £2..- As Also 5 Silver Spoons w^t:
7^{oz}: 10^{dwt}: wch. Were given away by the Widow.

W^m: Fairfield
David Cutler
Joseph Bradford

Suffolk Ss. By the Honble. Thomas Hutchinson Esqr. Judge of Prob^{ts} &c.
John Franklin Surviving Executor presented the abovewritten & made Oath
that it contains a true and perfect Inventory of the Estate of his Father Josiah
Franklin deced, so far as hath come to his Knowledge & that if more hereafter
appears he will cause the same to be added.
Boston Feb^{ry}: 12th: 1753.

T. Hutchinson

Exam^d:

Document 24

Executor's Account of Josiah Franklin's Estate, February 12, 1753

(*Suffolk Probate Records*, 49: 80-82)

Franklin's
Exec[rs]: Acco[t]:

Suffolk Ss: The Accoumpt of John Franklin Executor of
the Testamnet of his father Josiah Fraklin late of Boston deceas'd

The said Accomptant chargeth himself wth. the Goods & Chattels of the s[d]:
deced specify'd in a Inventory thereof by him exhibited into the Probate office
for the County aforesaid the 12th day of feb[y]:
A.D. 1753 Amounting to the Sum of . . . .　　　£59..12..2½

As also with Sundries since receiv'd viz[t]:

By Cash for Shop Rent . . . .　　　3..9..4

By D°. of Edw:[d] Malcom . . . .　　　10..5..6

By what the Goods sold for more than Appraisement . . . .　　　11..5..1

By five Silver Spoons not apprais'd given away
　　by the Widow valu'd at . . . .　　　3..6..8

By a High Candlestick valu'd at . . . .　　　8...

　　　　　　　　　　　　　　　　　Brought over　　£88..6..9½

By an old Copper not apprais'd w[t]: 280*l* . . at 1/2½ . . . . .　　　16..17..1½

　　　　　　　　　　　　　　　　　　　　　　　　£105..3..11

Debts outstanding viz:[t]

　　　from Edw[d]: Malcolm . . . . . .　　23..8..2
　　　from M[rs]. Dowres' Estate . . .　　6..13..4
　　　from W[m]. Homes . . . .　　6..2..8

　　　　　　　　　　　　　　　　£36..4..2

And the said Accomptant prays Allowance as follows vizt:

Paid for Mourning to the Widow	£36..12..2
Paid for 2 Gall^s. Wine for the Funeral	2..10..0
Paid Expences on the Porters	4..11
Paid for Gloves for the Funeral	15..9..6
Paid John Poluton for Bells, Pall &c	6..2
Paid for the Coffin	4......
Paid for Sundrey Necessaries for the Widow	9..8..4
Paid M^r. Venner for the Post boy	13..6
Paid M^r. Dupee Collector for Rates	2..7...
Paid M^r. Thornton	1..3...
Paid M^{rs}. Sutton Interest at Sundry times	20......
Paid Eben^r.. Storer his Ballance	23..18..6
Paid Deacon Grant towards repairing the common Shore .	1.....

Old Tenour	£123..8..11
in Lawfull Money	16..9..4

Paid John Beechum	27.....
Paid James Gordon	19..7..5
Paid D^o. as the deceas'd was Surety for Ja^s. Davenport	14..4...
Paid Expences at taking the Inventory	4..5
Paid for proving the Will, recording copy &c.	17..4
Paid advertizing the Sale of the Goods	6..8
Paid Expences at the Sale	7..4½
Paid for a Warrant for Appraisement	2..6

	£78..19..½

Paid the Appraiser 18/ their Expences 3/9½	£1..1..9½
Paid the Cryer 2/ for Paper 1/	..3....
Paid advertising the House to be sold	.11..4
Paid M^r. Nichols for Sale of the Goods	3..1..6
Paid James Gordon Interest	1..4..9¼
Paid Mrs Sutton more Interest	2..2..8
Paid Principal & Interest of Capt: Turell Bond	
wth. the difference of money	21..8..8

P^d: Principal & Interest & d°: on
        Benj. Bass his Bond . . . .   24.....6
P^d: for recording the Inventory Oaths &c. . . . .   .3..8
Paid for drawing, examining & allowing this Acc^t: &c . . . . .   7..2
P^d: for drawing List of Debts, examining & Certificate . . .   .3..
P:^d the Widow funeral Charges . . . .   6.10..9

                Brought over   78..19..½

      Save Errors φ John Franklin   £139..17..10¼

Suffolk Ss: By the Honble: Thos: Hutchinson Esqr: Judge of the Probate &c
John Franklin Executor presented the within written & made Oath that it
contains a just & true Acc^t: of his Administration of the Estate of Josiah
Franklin deceas'd so far as he hath proceeded therein & produc'd Receipts &
Vouchers for the several Payments therein mention'd, wch I do allow & approve
of. Boston Feb^y: 25. 1754

Exam^d: φ John Shirley Reg^r: T Hutchinson

Document 25
John Franklin's Petition to the Massachusetts Superior Court,
February 26, 1754
(By Permission of the Boston Public Libray)

Province of the ⎫   To the Hono^ble. his Majesty's Justices of
Massachusetts Bay ⎬   the Superior Court of Judicature
Suffolk Ss: ⎭   Sitting at Boston Feb^Y. 26, 1754

Humbly Shewn John Franklin Executor of the Testament of Josiah Franklin late of Boston in the County of Suffolk Tallow-Chandler dec^d. That the said Deceased's ˄ ^Personal Estate is Insufficient to pay his Just Debts the Sum of One hundred & Sixty Six Pounds Seven Shillings & eleven pence as appears by a Certificate from the Probate office herewith presented___
    Your Petit^r. therefore humbly Prays this Hono^ble. Court /before whom the Premes are Cognizable/ to authorize & Impower him in his said Capacity to sell & dispose of the said testator's Real Estate, consisting of a house & Land in Union Street in Boston aforsiad for the payment of his Just Debts ___

            And as in Duty Bond &c ---

                John Franklin

Suffolk Ss:

These Certify all Concerned That John Franklin ~~admin~~ ^Executor of the ~~Estate~~ ^Testament of his Father Josiah Franklin Late of Boston in the County of Suffolk Tallow Chandler deceaced hath Exhibited an Inventory of his said Testator's Estate into the Probate office for the County aforesaid amounting to the Sum of Three hundred & fifty eight Pounds ten Shillings & seven pence whereof in Real Estate /consisting of a House & Land in Union Street/ Two hundred & fifty three Pounds Six Shillings & eight pence Personal, One hundred & five Pounds three Shillings & eleven pence/ besides the Sum of Thirty Six Pounds four Shillings & two pence outstanding Debts/ and I do further Certify That the said Executor has rendered an accompt of his Said Adminacon upon Oath whereby it appears that he has paid and is allowed the Sum of One hundred & thirty nine Pounds Seventeen Shillings & ten pence and that there is still Due & owing him the Said Deceaced's Estate the Sum of One hundred & thirty One Pounds f[ou]rteen Shillings as ꝑ a list of Debts filed in the Probate office --

            Dated at Boston The 25th. Day
            of February Anno Domi 1754

                John Shirley Reg^r:

DOCUMENT 26

John Franklin's Indenture of Sale to William Holmes, April 15, 1754
(*Suffolk Deeds*, 85: 64)

Franklyn to
Homes

This Indenture made the fifteenth day of April One thousand seven hundred
and fifty four in the twenty seventh year of His Majesty's Reign Between John
Franklyn of Boston in the County of Suffolk and province of the Massachusetts
Bay in New England merchant surviving Excor of the last will and Testament
of his Father Josiah Franklyn late of said Boston Tallow Chandler deceased on
the one part and William Homes of Boston aforesaid Goldsmith of the other
part Whereas the said John Franklyn Excor as aforesaid in answer to his
petition Preferred to his Majesty's Justices of the Superiour Court of Judicature
held at Boston aforesaid on the Third Tuesday of February last was by Virtue
of an Order from the said Court then Obtained Authorized and Impowered to
make Sale of the deceased's House and Land hereafter described for the
payment of his just Debts and to Execute a good Deed or Deeds in the Law for
Conveyance thereof the petitioner posting up notifications thirty days before
the sale thereof as by the said Order (reference thereunto being had) will more
fully appear and Whereas the said John Franklyn has observed the directions of
the Law by posting up Notifications thirty days before the sale and on the ninth
day of april Current caused the said House and Land to be sold at the public
Vendue to the highest bidder and the said Willm: Homes being the highest
bidder and giving most therefor the same was sold to him This Indenture
therefore Witnesseth that the said John Franklyn Excor as aforesaid in
pursuance and by Virtue of the aforerecited Order of Court for and in
consideration of the Sum of One hundred and Eighty eight pounds thirteen
shillings and four pence lawful money to him in hand paid before the Ensealing
hereof by the said William Homes the receipt whereof is hereby acknowledged
(the same to be apply'd towards the payment of the said deced's just Debts)
Hath granted bargained sold aliened conveyed & confirmed and by these
presents doth fully and absolutely grant bargain sell aliene convey and confirm
unto him the said William Homes All that certain peice or parcel of Land
situate lying and being in Boston aforesaid and is butted & bounded as follows
vizt: Easterly on Union Street there measuring thirty nine feet Northerly on
Hanover Street there measuring ninety four feet Westerly on Land of M^rs:
Dorothy Carnes there measuring twenty four feet and southerly on land of
Jeremiah Bumstead deced there measuring Eighty seven feet or however
otherwise the same is bounded or measures Together with the Dwelling

Houses Edifices and Buildings thereon Rights Members and appurces thereto belonging To Have and To Hold said Granted Land Edifices Buildings Premisses Privileges and Appurces unto the said William Homes his Heirs and Assigns to his and their only proper use benefit and behoof forever And the said John Franklyn Excor as aforesaid for himself his Heirs Excors & Admors doth hereby covenant with the said William Homes his Heirs and Assigns in manner following That is to say, that the said Josiah Franklyn deceased in his life time and at the time of his Death was lawfully seized and possessed of the said granted Premisses in his own proper Right as of a good perfect and absolute Estate of Inheritance in Fee Simple And that by Virtue of the aforerecited Order of the Superiour Court he the said John Franklyn Excor as aforesaid hath in himself full power to Grant Sell and Convey the same in manner as aforesaid the premisses being free and clear of and from all manner of former or other gifts Grants Bargains Sales Leases Releases Mortgages Wills Entails Joyntures Dowries Judgments Executions and Incumbrances whatsoever and Further that he the said John Franklyn his Heirs Excors and Admors in his capacity aforesaid shall & will warrant and Defend all the said granted premisses with the appurces un to the said William Homes his Heirs and Assigns forever against the lawful Claims & Demands of all persons whomsoever In Witness whereof the said John Franklyn Excor as aforesaid hath hereunto set his hand and seal the day and year first aforewritten John Franklyn and a Seal Signed Sealed and Delievered in presence of us John Barrett Jonathan Williams Suffolk Ss: Boston April 26th: 1754 M^r: John Franklyn Excor &c: personally appeared and acknowledged the aforewritten Instrument to be his free Act and Deed Before me John Phillips Just^o: Pacis April 27th: 1754 Received and accordingly Entred and Examined ___

                    ℘ Ezek^l. Goldthwait Reg^r..

DOCUMENT 27

Probate Letter, John Franklin's Will and Codicil,
January 22 & 24, 1756
(*Suffolk Probate Records*, 51: 93-98)

Probate
  of
John Franklin

Thomas Hutchinson Esqr. Commiss^d ~ by the Governour by & with the Advice, & Consent of the Council for the Province of the Massachusetss Bay to be Judges of the Probate of Wills, & for granting Letters of Administr^n~ on the Estates of Persons dec^d~ having Goods, Chattels, Rights, or Credits, in the County of Suffolk, within the Province afores^d.~ To all unto whom these Presents shall come ~ Greeting ~ Know ye that upon the Day of the Date hereof before me at Boston, in the County afores^d. The Will ⌃ $^{\&\ Codicill\ to\ the\ will}$ of John Franklin late of Boston in the County afores^d. Tallow Chandler dec^d. to these Presents annexed, was proved, approved, & allowed, who having while he lived, & at the time of his Death, Goods, Chattels, Rights, or Credits, in the County afores^d., And the Probate of the s^d. Will, & Powers of committing Administr^n_ of all & singular the Goods, Chattels, Rights, & Credits of the the s^d. dec^d. by Vertue thereof appertaining unto me. The Administration of all & Singular y^e. Goods, Chattels, Rights, & Credits of the s^d. dec^d. & his Will in any manner concerning is hereby committed unto Jonathan Williams, and Tuthill Hubbart, Exec^{rs}_ in the same Will named well & faithfully to execute the s^d_ will, & to administer the Estate of the s^d. dec^d. according thereunto: And to make a true & perfect Inventory of all & singular the Goods, Chattels, Rights, & Credits of the s^d. dec^d. & to exhibit the same into the Registry of the Court of Probate, for the County afores^d. at or before the sixth day of May next ensuing: And also to render a plain and true Accompt of their s^d_ Administration upon Oath. In Testimony whereof I have hereunto set my Hand, & the Seal of the s^d_ Court of Probate as Dated at Boston the Sixth day of Feb^y_ Anno Domini 1756.

Jno. Cotton Reg^r.   T Hutchinson

Franklin's
Will_

In the Name of God, Amen. This Twenty Second day of January, Anno Domini, One Thousand Seven Hundred, & Fifty Six, & in the Twenty Ninth Year of His Majesty's Reign. I John Franklin of Boston in the County of

Suffolk, and Province of the Massachusetts Bay, in New England Tallow Chandler, do make this my last Will, & Testament, in manner following. First, & principally, I resign, & recommend my Soul into the Hands of Almighty God, & my Body to the Earth, Trusting, & hoping in the Infinite Mercies of God for a Resurrection to Eternal Life, thro' the Influence of Our Lord Jesus Christ. As to my wordly Estate which it has pleased God to bestow on me, I will, & dispose thereof as follows. Imprimis, I Will that all my just Debts, & Funeral Charges, be paid in due time, by my Executors hereafter named. Item, whereas I have agreed with the other Proprietors of German Town so called in Braintree, in the County of Suffolk afores^d. upon a Partition of the same among Ourselves; My will is that Partition be made thereof, & I do hereby impower Jonathan Williams, & Tuthill Hubbart both of Boston afores^d. Merchants to make & execute proper Deeds of Partition with the other Proprietors, & will that my Interest in the s^d. German Town shall be as much bound by the Agreements, & Covenants, & Deeds made by the s^d. Jonathan Williams, & Tuthill Hubbart, as if they were made, & executed by me in my Life time. Item, I hereby give, & devise, to my well beloved Wife Elizabeth, the one Third part of my Real & personal Estate, to be to her, & her Heirs & assigns forever agreable to my Marriage Covenants with her: And whereas by a Marriage Contract, I was also to receive of Joseph Gooch Esqr. Trustee for my s^d. Wife. The Sum of Two hundred & Fifty Pounds in Bills of the new Tenor, great part of w^ch., I have ⌃ ^rec.& hereby remit whatsoever remains unpaid of the same Sum. Item, I give to my well beloved Wife Elizabeth, during her natural Life, the use of all the Utensils belonging to my Business /That is to say/ the Copper, the Iron Furnace, & other Implements and also the use of my Negro named Cæsar; and after her Decease I give all the same to my Kinsman Peter Franklin Macom, & James Barker equally between them. Item, I give to my s^d. Wife a Suit of Mourning. Item, I give to my Kinsman Peter Franklin Macom a Suit of Cloaths, but not mourning. Item, The Remainder of my Real, & personal Estate, I give to my Brethren, & Sisters equally, & to the legal Representatives of such of them as are dece^d. Lastly I nominate, & appoint Jonathan Williams, & Tuthill Hubbart of Boston afores^d. Merchants, Executors of this my last will, & Testament, Revoking all former Will, & Wills, by me heretofore made. In Witness whereof, I have hereunto set my Hand, & Seal, the day & Year above written.

John Franklin & a Seal.

Signed Sealed Published, & declared by---
the s^d. John Franklin, to be his last Will
& Testament in Presence of us, who Subscribed
our Names in the Testaters Presence
John Perkins, Silv. Gardiner, Gillan Tylor.

Codicil   A Codicil to the foregoing Will.

I John Franklin of Boston, in the County of Suffolk Tallow Chandler, This Twenty fourth day of January Anno Domini One Thousand, Seven Hundred, & Fifty Six. Do make, & publish this my Codicil to my last Will, & Testament, in manner following /That is to say/ I give to M^r. Jonathan Williams, & Tuthill Hubbart whom I have appointed my Executors, to each of them the Sum of Thirteen Pounds Six Shillings, & Eight pence for their Trouble. And whereas, I have give to my wife the use of my Negro named Cæsar, during her natural Life, I now give my Executors Liberty, to dispose of him if he behaves ill, & my s^d. wife to have the use of the Money, & at her decease to be divided equally between my kinsmen, Peter Franklin Macom, James Barker, & John Macom. I give to my Sister Jane Macom 1 pair of Silver Canns, with my Arms upon them. I give to my Son in Law Thomas Hubbart, one Silver Porringer. I give to my Daughter in Law Susannah Hubbart, one Silver Porringer. I give to my Daughter in Law Elizabeth Hubbart my Two Volumes of Chamber's Dictionary, & Lord Bacon's Works, in three Volumes. I give to my Niece Grace Williams, the Wife of M^r. Jonathan Williams my Silver Salver. I give to my well beloved Wife, my large Picture, likewise my Brother Benjamin Franklin's Picture during her natural Life & after her Decease, I give my Picture to William Franklin of Philadelphia, the Son of my Brother Benjamin Franklin Esqr; And I give my Brother Benjamin Franklin's Picture to James Franklin of Newport the Son of my Brother James Franklin dec^d. I give to my Maid Elizabeth Clements, the Sum of Two Pounds, Thirteen Shillings, & four pence. In Witness whereof, I have hereunto sett my Hand, & Seal, the Day, & Year above written.

John Franklin
& a Seal.

Signed Sealed, & Declared by the)
s^d. John Franklin as a Codicil to   )
his Will, & Testament in Presence }
of Jos: Edwards, Thos. Leverett   )
   Dudson Kilcup --------   )

Sulffok Ss: The within written Codicil being Presented for Probate. Thomas Leverett, & Dudson Kilcup made Oath that they saw John Franklin the Subscriber to this Instrument, Sign, & Seal, & heard him publish, & declare the same to be his last Will, & Testament, & that when he so did, he was of sound disposing Mind, & Memory according to these Deponents best discerning, & that they together with Joseph Edwards now absent, Set to their Hands, as witnesses thereof, in the s^d. Testators Presence. Boston Feby. 6th. 1756.

T Hutchinson

Ex^d. ℗ Jno. Cotton Reg^r.

Document 28

Price Lists of Candles, Tallow, Soap, and Others[1]

	Candles doz. lbs.		Tallow cwt.			Candles doz. lbs.		Tallow cwt.	
	s.	d.	s.	d.		s.	d.	s.	d.
1583-4	3	1¾			1603-4	3	3½		
1584-5	3	4			1604-5	4	4½		
1585-6	3	5	24	0	1605-6	3	10¾		
1586-7	3	2¼			1606-7	3	9½		
1587-8	3	5½	34	6	1607-8	3	11¼		
1588-9	3	11½	30	0	1608-9	4	0¼		
1589-90	3	10¾	29	0	1609-10	4	2½		
1590-1	3	7½	32	8	1610-1	4	10		
1591-2	3	9½			1611-2	4	1		
1592-3	3	7¼			1612-3	4	3		
1583-1592 Average	3	6½	30	0	1603-1612 Average	4	0¾		
1593-4	3	8¼	33	4	1613-4	4	2		
1594-5	3	0¾	30	0	1614-5	4	8		
1595-6	3	8½			1615-6	4	11		
1596-7	3	7½	35	6½	1616-7	4	7¼		
1597-8	4	0¾	37	4	1617-8	4	4¼		
1598-9	4	2¾			1618-9	4	6		
1599-1600	4	1	37	0	1619-20	4	6	37	4
1600-1	4	7½	39	1	1620-1	4	9¾	37	4
1601-2	4	5			1621-2	4	6½		
1602-3	3	7			1622-3	4	1¾	28	0
1593-1602 Average	3	11	35	4½	1613-1622 Average	4	6¼	34	3

[1]Unless indicated otherwise, all data were quoted from James E. Thorold Rogers, ed., *A History of Agriculture and Prices in England* (1887; Vaduz, Japan, 1963), vol. 5, pp. 398-404, 580-84, 740, 747, 751; vol. 7, pp. 312-19, 407-09, 458-92, 592-93, 598-99. He drew information mainly from the purchasing records of such corporations as London, Oxford, Cambridge, York, and so forth. Many prices, therefore, were of wholesale and may vary in other regions in England.

	Candles doz. lbs.		Tallow cwt.	
	s.	d.	s.	d.
1623-4	4	2	28	0
1624-5	5	0½	35	1
1625-6	4	0½	41	11
1626-7	4	4¾	38	7
1627-8	4	7½	37	0
1628-9	4	6¼	38	5
1629-30	4	5	37	4
1630-1	4	3¾	41	0
1631-2	4	2¾		
1632-3	4	3¼		
1623-1632 Average	4	4¾	37	2
1633-4	4	7¼	37	4
1634-5	5	1½		
1635-6	5	1¾		
1636-7	4	8	28	0
1637-8	4	10	33	8
1638-9	4	11	28	0
1639-40	4	10½		
1640-1	4	11½		
1641-2	5	6		
1642-3	4	11½		
1633-1642 Average	4	11½	31	9
1643-4	4	7		
1644-5	4	7	23	4
1645-6	4	7½	23	4
1646-7	6	0¼	23	4
1647-8	6	5¾	28	0
1648-9	6	0½	28	0
1649-50	6	5½	28	0
1650-1	5	10	28	0
1651-2	5	9¾	28	0
1652-3	5	3¼	28	0
1643-1652 Average	5	8	26	3

	Candles doz. lbs.		Tallow cwt.	
	s.	d.	s.	d.
1653-4	5	2½	28	0
1654-5	4	5¾	23	4
1655-6	4	2¼	23	4
1656-7	4	6¼	23	4
1657-8	5	3½	23	4
1658-9	5	2½	28	0
1659-60	4	11	26	10
1660-1	5	8¼	28	0
1661-2	5	7¼	28	0
1662-3	5	7	28	0
1653-1662 Average	5	0¾	26	0½
1663-4	5	8	28	0
1664-5	5	9	28	0
1665-6	5	6		
1666-7	5	6	28	0
1667-8	5	0½	25	8
1668-9	4	10½	23	4
1669-70	5	0	23	4
1670-1	5	5	22	2
1671-2	4	8½	22	2
1672-3	4	6	22	2
1663-1672 Average	5	2¼	24	9
1673-4	4	9	25	8
1674-5	5	4	25	8
1675-6	5	3¾	25	8
1676-7	5	9¼	25	8
1677-8	5	6¼	25	8
1678-9	5	3	25	8
1679-80	5	4	25	8
1680-1	4	10	25	8
1681-2	4	8	25	8
1682-3	4	4½	25	8
1673-1682				

Average 5 1¼ 25 8

Period	Candles doz. lbs. s.	d.	Tallow cwt. s.	d.	Year	Wax Candles s.	d. @ lb.
					1703	1	7½
					1704	1	6
1683-4	4	7½	22	2	1705	1	7
1684-5	5	4	25	8	1706	1	7
1685-6	5	0	25	8	1707	1	7
1686-7	4	9	25	8	1708	1	7
1687-8	4	5¾	25	8	1709	1	7
1688-9	4	7¾	23	4	1710	1	11
1689-90	4	10½	25	8	1711	2	3
1690-1	4	6¾	25	8	1712	2	3
1691-2	4	5	25	8	1713	2	3
1692-3	4	9¾	25	8	1714	2	3
					1715	2	3
					1716	2	3
1683-92					1717	2	3
Average	4	9	25	1	1718	2	2
					1719		
					1720	2	2
1693-4	5	9½	25	8	1721	2	3
1694-5	6	5	25	8	1722	1	6
1695-6	5	11	24	6	1723		
1696-7	5	7	25	8	1724		
1697-8	5	5¾	25	8	1725	2	5
1698-9	5	3¾	28	0	1726	2	6
1699-1700	5	4	29	2	1727		
1700-1	5	2	28	0	1728	2	7
1701-2	5	8			1729	2	3
1702-3	5	4			1730	2	3
					1731	2	4
					1732	2	3
1693-1702					1733		
Average	5	7¼	26	6½	1734	2	2
					1735		
					1736		
1583-1703					1737		
Average	4	8¾	28	5¾	1738		
					1739		
					1740		
					1741		
					1742		
					1743		
					1744		
					1745		
					1746		

1747	2	4		
1748	2	4		
1749	2	4		

Wax Candles

	s.	d.	@ lb.
1750	2	6	
1751	2	6	
1752	2	6	
1753	2	6	
1754	2	4	
1755	2	6	
1756	2	6	
1757	2	6	
1758	2	6	
1759	2	6	
1760	2	6	
1761	2	6	
1762	2	6	
1763	2	6	
1764	2	6	
1765	2	6	
1766	2	6	
1767	2	6	
1768	2	6	
1769	2	6	
1770	2	6	

Spermaceti Candles

	s.	d.	
1771	2	2	@ lb.
1772	28		@ doz.
1773	24		@ doz.
1774	24		@ doz.
1775	24		@ doz.
1776	27		@ doz.
1777	34		@ doz.
1779	3	4	@ lb.
1780	42		@ doz.

Rush-lights

	s.	d.	
1745	5	6	@ doz.
1746		5½	@ lb.
1747	5	4	@ lb.
1748		6	@ lb.
1749	5		@ doz.
1751		6	@ lb.

1753	4	10	@ doz.
1756	6	4	@ doz.
1784	7	6	@ doz.
1787	8	6	@ doz.
1790	7	8	@ doz.

Tallow

1712   Tallow, stone(8 lb.)
     @ 2/3. to 2/5.
     Country stone(14 lb.)
     @ 4/.
1714   Tallow, cwt. @ 36 to 42/.

Wax

1728   White wax
     3/. @ lb.
1735   English wax
     110/. @ C[wt]
     Foreign wax
     80/. @ C
1739   Old wax
     125/. @ C
     New wax
     120/. @ C
     Long wax
     2/9. @ lb.

Special Candles

1715   10 doz. mould candles
     8/3½. @ doz.
1748   6 lb. mould candles
     6/4. @ doz.
1749   ½ doz. mould candles
     6/.   @ doz.
1755   1 doz. mould candles
     7/.   @ doz.
1760   Chapel candles
     3/.   @ lb.
1761   Small candles
     6/.   @ lb.
1762   Yellow wax candles
     2/6. @ lb.
1766   Rolled wax candles

      3/. @ lb.
1768 3 lb. best mould
      candles
      /8. @ lb.
      6 best mould candles
      /8. @ lb.
1770 4 lb. stable candles
      /7. @ lb.
1772 1 doz. best mould
      candles
      8/6. @ doz.
1773 1 doz. best mould
      candles
      8/. @ doz.
1785 1 lb. of sixteens 3/3.

White and Yellow Candles

1 set white candles
      1784 @ 7/.
      1785 @ 6/2.
      1786 @ 6/2.
      1787 @ 6/2.
      1788 @ 6/2.

1 set yellow candles
      1784 @ 3/6.
      1785 @ 3/6.
      1786 @ 3/6.
      1787 @ 3/6.
      1788 @ 3/6.

Candlesticks

1708 Brass candlestick @ 3/9.
1716 Brass hand candlestick
      1/10.
1722 2 new desk candlesticks
      36/.
1748 2 tin candlesticks 1/4.
1760 Strong flat candlesticks
      3/.
1762 12 brass candlesticks
      @ 1/6.
1764 Tin candlestick /8.
1772 Iron Candlesticks 2/6.
1775 4 candlesticks 3/4.
1782 2 pairs candlesticks £5
      5/.
1784 Iron double candlestick
      /8.

Candle-Wick and Camphor

1602 Candle-wick 5s. @ doz.
1603 Candle-wick 7s. @ doz.
1785 2 oz. camphor @ /9.
1790 3 oz. Camphor 2/.
1791 ½ oz. Camphor @ /8.

Olive Oil

1692 3/8. to 4/6. @ gallon
1693 2/8. to 4/4. @ gallon

Soap

1598 10s. @ firkin of 64 lbs.
1633 25s. @ firkin of 64 lbs.
1653 10s. @ firkin of 64 lbs.
1696 22s. @ firkin of 64 lbs.
1705 12 firkins £12 12/.
1706 12 firkins £12
1708 12 firkins £11
1710 6 firkins £6
1711 doz. @ 4/8½.
1745 6/. @ st.
1746 5/8. @ st.
      /5. @ lb.
1747 6/10.@ st.
1748 6/. @ st.
      6/10.@ st.
1749 5/10.@ st.
      /10.@ lb.
1751 /6. @ lb.
1753 5/10.@ st.
1754 6/4. @ st.
1755 6/4. @ st.
      6/10.@ st.
1756 6/10.@ st.
1759 6/4. @ st.
1761 6/10.@ st.
1762 6/10.@ St.
      /6. @ lb.
1766 7/6. @ st.
1767 7/. @ st.
      7/6. @ st.
1768 7/. @ st.
1770 7/. @ st.

1771	7/. @ st.	
1779	7/6. @ st.	
1784	/8. @ lb.	
	2/. 3 lb.	
1789	8/2. @ st.	
1790	8/2. @ st.	
1791	8/6. @ st.	

Other Soaps

Sweet or black soap
1633-98   4s. to 7s. @ doz.

Windsor soap
1780   2s. @ lb.

Yellow soap
1784   8s.9d. @ st.
1792   8s.6d. @ st.
1793   8s.6d. @ st.

Soft Soap

1709   85/.   barrel
1745   /11. 2 lb.
1793   /8. @ lb.

White soap
1785   2s. 4½d. for 3½ lb.
1787   10s.3d. @ st.

Hard Soap

1747	/6. @ lb.
1784	8/9. @ st.
1785	8/6. @ st.
1786	9/4. @ st.
1787	8/9. @ st
1788	8/2. @ st.

Castile Soap

1715	57s. @ cwt.
1745	6d. @ lb.
1746	6d. @ lb.
1747	6d. @ lb.
1748	6d. @ lb.
1749	6d. @ lb.
1753	7s. @ doz.
1755	2s. @ st.
1756	7s. @ st.
1759	7s. @ st.
1761	7s. @ st.
1762	7s. @ st.
1766	7s.4d. @ st.

Ball Soap

1745	1/10. 7 lb.
1766	/6½. 1 lb.
1771	7/. @ st.
1772	7/6. @ st.
1773	7/6. @ st.
1775	6/. @ st
1776	7/. @ st.
1778	7/. @ st.
1779	/7. @ lb.
	7/6. @ st.
1780	7/6. @ st.
1781	7/6. @ st.
1782	7/7. @ st.
1783	7/6. @ st.
1784	8/2. @ st.
1785	8/2. @ st.

	8s. @ st.
	6½d. @ lb.
1767	8s. @ st.
1768	8s. @ st.
1770	7s. @ st.
1771	7s. @ st.
1772	8s. @ st.
1773	8s. @ st.
1775	8s. @ st.
1776	8s.6d. @ st.
1778	9s. @ st.
1779	9s. @ st.
1780	9s. @ st.
1781	9s. @ st.
1782	9s. @ st.

1783	9s.6d. @ st.
1784	9s.9d. @ st.
1785	9s.6d. @ st.
	1s.4d. @ lb.
1786	10s. @ st.
1787	10s. @ st.
1788	9s.6d. @ st.
1789	9s.6d. @ st.
1790	9s.6d. @ st.
1791	9s.6d. @ st.
1792	10s. @ st.

Lime and Ashes

1715	Lime, 6d. @ bushel
1717	Lime, 6d. @ bushel
1786	4 load ashes 6s.
1791	1 load ashes 1s.

Silk and Other Fabric

Velvet   23s. 11¼ d. 1583-1702

Satin   10s. to 17s./yard
17s.(green) in 1608
20s.(white) in 1629
16s. 6d. (figured) in 1633
13s.(blue wrought) in 1649
16s. 8d. (scarlet) in 1651

Sarsnet   8s./yard in 1610
9s. 4d. in 1649
8s. 6d. in 1668
4s. 6d. in 1697

Taffeta   10s./yard in 1584
14s./yard in 1593
15s./yard in 1608

Tabby   12s. in 1650

Grogram   11s. 6d. in 1393
10s. 6d. in 1615
17s. 4d. in 1620
9s. 6d. in 1631

12s. 6d. in 1631

Ras de Cypre   8s. in 1649

Drap de Berry   14s. in 1649
11s. in 1688

Satanetto   2s. 6d. in 1686

Flower silk   8s. 6d. in 1686

Silk for breeches 10s.in 1687

Silk fringe
2s. 1½ d./ounce in 1662

Stamell   17s. 6d. in 1617

Plush   23s./yard in 1629
13s. 6d./yard in 1630
11s./yard in 1630
10s./yard in 1631

Damask   16s. 4d. in 1662

Figured damask   8s.in 1699

Rich tissue damask
12s. 6d. in 1700

Lutestring   13s.   in 1650
5s.   in 1667
6s. 6d. in 1701

Mantua (plain)   7s. in 1702

Mantua (yellow) 7s. 6d. in 1700

Mantua (black and white)
7s. 9d. in 1700
6s. 6d. in 1702

mantua (striped)7s. 6d. in 1702

Altar-cloth of 16¾ yards of
purple velvet, two yards of
Watchet damask (watered
silk), 15½ yards of fustian
lining, an ounce of Venice
gold, an ounce of silver

thread, and half an ounce of
gold pearl. £17 9s. in 1605

Bologna silk lb. (16 oz.)
26/6. to 27/6. in 1709

China silk lb. (16 oz.)
21/. to 22/. in 1709

Italian silk lb. (16 oz.)
27/6. to 28/6. in 1720

Silk Product

1707 Black silk waistcoat 46/.
1709 2 pair silk garters @
5/4½.
2 black silk laces @ /9.
1716 2 silk handkerchiefs @
2/9.
1717 silk hood 26/.
1720 2 silk strings 25/.
1724 New silk hood 38/9.
1743 Pair silk garters 1/9.
1756 Pair braided silk 5/.
Pair blue satin 4/.
1761 Pair black silk garters
1/6.
1762 Stein of silk /3½.
1763 2 pair silk puffs 2/.
1766 Pair black satin shoes
10/.
1767 Pair green satin heels
7/6.
Pair white satin heels
11/.
1770 3 pair super white silk
36/.
3 pair men's fine silk
and worsted twisted
19/6.

1773 Pair silk shoes for lady
/6.
1774 Velvet cap 14/.
Pair china white silk
hose 15/.
1776 Pair men's white china
pattern rib'd silk
stockings 14/6.
3 pair men's 24 stout
white silk stockings
workt clocks @ 15/.
1777 2 pair men's random china
silk stockings
workt clocks @ 14/.
1778 6 pair men's best super
fine white china
silk hose @ 15/.
1778 Pair large men's col^d
pattern ribbed
silk hose @ 16/.
1780 silk hat 12/.
Set of green silk spring
curtains £3 13/6.
1782 Pair black silk women's
shoes 12/.
1783 An embroidered waistcoat
on satin £2
1785 2 pair large white china
silk hose 31/6.
Silk & muslin chequed
waistcoat 18/.
1786 Silk & muslin tissue
waistcoat 18/.
1789 6 pair women's china
silk hose £3 3/.
1790 White satin cloak 66/.
Pair black silk hose 14/.
1792 3 pair ladies white
silk hose @ 10/.

From the *Boston Gazette*

	1719 January 11	1719 December 21	1720 April 11 to May 23
Beeswax		2s. 2d. @ lb.	2s. 2d. @ lb.
Bayberry wax	16d.@ lb.	20d. @ lb.	16d.@ lb.
One tallow candle			10d.

From Josiah Franklin's Bill of 1711
(See Document 9)

Tallow Candles

1	9d.
2	1s. 6d.
3	2s. 3d.
4	3s.
5	3s. 4d. to 3s. 9d.
½ doz.	4s. to 4s. 6d.

Chronology

1657   December 23, born at Ecton, Northamptonshire, England, youngest son of Thomas Franklin 2nd and Jane White Franklin.

1662   October 30, Jane Franklin (b.1617) dies.

1666   Thomas Franklin 2nd moves out from the eldest son Thomas 3rd's house and stays with another son John, a dyer at Banbury, Oxfordshire, who also takes Josiah as his apprentice.

1667   August 15, Abiah Folger born, daughter of Peter and Mary Morrils (Morrill) Folger of Nantucket, New England.

c.1677 Marries Ann(e) Child of Ecton, daughter of Robert Child.

1678   March 2, first daughter Elizabeth born at Ecton.

1681   May 16, first son Samuel born at Banbury.

1682   March 21, Thomas Franklin 2nd dies at Banbury.
        Brother John marries Ann Jeffs of Marton, Warwickshire.
        Brother Benjamin (The Elder) comes to Banbury from London.

1683   May 25, Hannah born at Banbury.
        July, Josiah, Anne, and their three children leave Banbury, via London, for New England.
        October, arrive in Boston.
        November 23, Benjamin Franklin the Elder marries Hannah Welles, daughter of the dissenting minister Samuel Welles of Banbury.

1685   August 23, son Josiah born.
        Rents a tenement house in Milk Street.
        Receives baptism and applies for membership at the South Church.

1687   January 5, daughter Ann(e) born.

1688   February 6-11, son Joseph I born and dies.
        August 19, Abiah Folger receives baptism at the South Church.

1689   June 30 - July 15, Joseph II born and dies.
        July 9, wife Ann(e) dies, buried in Granary Burial Ground.
        November 25, marries Abiah Folger, at the South Church by the Reverend Samuel Willard.

1690   December 7, John born, first son with Abiah.

1691   April 27, Town Meeting proves his proposed 8 by 8-foot "lean-to"
        to be added to his tenement.
        November 22, Peter born.

1694   February 4, Josiah and Abiah granted full membership of the South Church
        and allowed to partake communion.
        September 26, Mary born.

1696   November 18, signs a petition against restrictions on building timber houses
        in Boston.

1697   February 4, James born.
        May 4, elected tithingman.

1698   The Blue Ball made and later displayed at a rented shop close to the corner of
        School and High (or Cornhill, later Washington) Streets.

1699   July 9, Sarah born.

1701   March 10, elected market clerk.
        September 20, Ebenezer born.

1703   February 5, Ebenezer drowns.
        March 8, elected constable.
        December 7, Thomas born.

1705   October, joint ventures with Peter Folger Jr. and John Folger to make and sell
        rush-candles.

1706   January 17, Benjamin born.
        August 17, Thomas dies.

1707   January 8, Elizabeth, at 29, marries Joseph Berry, shipmaster.

1708   August 8, Lydia born.
        September 7, hosts his first neighborhood prayer meeting at his tenement
        house.

1712   January 25, buys from Peter Sergeant a property for £320 (Old Tenor) at the
        south-west corner of Union and Hanover Streets; February 8, closes the
        purchase by mortgaging the house to borrow £250 from Simeon Stoddard.
        March 27, last child Jane born.

1715   March 14, elected tithingman.

October 10, Benjamin Franklin the Elder arrives.

1717   June 21, Benjamin Franklin the Elder begins to write the family history at Josiah's house.

1719   April 17, fails to be chosen as deacon of the South Church.

1720   June 7, borrows more than £200 from James Bowdoin to help son James, printer.

1721   March 14, elected tithingman.
          August 7, James's *New-England Courant* launched.

1722   June 12-July 7, James arrested and put in jail.
          July 1, Josiah's apprentice runs away.

1723   January 29, mortgages the house again to Hannah Clarke for £220.
          May 1, signs a bond, with James Davenport (husband of Sarah Franklin), to borrow £100 from Josiah Hobbs for one year.
          September 25, Benjamin runs away.
          Subscribes to Samuel Willard's *Compleat Body of Divinity* for two copies.

1724   May 1, unable to pay the bond to Josiah Hobbs on schedule; unable to provide financial support to set Benjamin up.

1725   April 1, the bond to Josiah Hobbs extended for another year.

1727   March 17, Benjamin Franklin the Elder dies, buried in Granary Burial Ground.

1729   February 13, sued by Josiah Hobbs for failure to pay the bond.

1736   Subscribes to Thomas Prince's *History of New England* for two copies.

1739   August 9, pays back his loans in full to Hannah Clarke Willard and the mortgage is canceled.

1744   October 20, last will prepared.

1745   January 16, dies at 88.
          January 17, John Draper'e eulogy appears in *Boston News-Letter*.

1750   August 7, will probated.

1752   May 8, Abiah dies at 84.

1754   February 26, son John petitions the Massachusetts Superior Court to grant
        permission to sell Josiah's house in order to redeem his parents' debts.
        April 15, house sold to William Holmes, son of Mary Franklin and Robert
        Holmes, for £188. 13s. 4d.(Lawful Money)

1810   December 29, the old tenement house in Milk Street destroyed by fire.

1827   A group of Bostonians repairs the Franklin family tomb in the Granary
        Burial Ground and erects a monument.

1858   November 10, the house at the corner of Union Street demolished to widen
        the street. Only the Blue Ball has been preserved.

INDEX

www.ingramcontent.com/pod-product-compliance
Lightning Source LLC
Chambersburg PA
CBHW080927100426
42812CB00007B/2390